Industry 4.0 for Manufacturing Systems

The book highlights the importance of intelligent decision-making in advanced production systems and optimization of process parameters using fuzzy-based multi-criteria decision-making tools. It discusses the decision-making aspects of Industry 4.0 using machine learning and optimization techniques and helps in moving toward the digitalization of manufacturing systems. It further covers several important topics including the role of digital twins in advanced manufacturing processes, machine learning-based prediction of overall equipment effectiveness, intelligent quality control tools, and life cycle assessment models in Industry 4.0.

Key features:

- Presents a conceptual framework to measure the readiness of adopting Industry 4.0 in advanced manufacturing systems.
- Discusses the impact of smart manufacturing on sustainable development and integration of Industry 4.0 and additive manufacturing.
- Covers topics such as intelligent automation systems, machine learning-based preventive maintenance, and the Internet of Things-enabled additive manufacturing in Industry 4.0.
- Explains cyber-physical system integration with Industry 4.0 technologies, cyber-physical systems in industrial robotics, and green cyber-physical systems.
- Illustrates optimization of process parameters using fuzzy-based multi-criteria decision-making tools and life cycle assessment models in Industry 4.0.

This book is primarily written for senior undergraduates, graduate students, and academic researchers in the fields of industrial engineering, production engineering, mechanical engineering, supply chain management, and manufacturing engineering.

Industrial Engineering, Systems, and Management

Kapil Gupta, Jayakrishna Kandasamy

Industrial engineering, systems, and management is a broad term that includes theories, principles, concepts, tools, and techniques. This series will cover industrial engineering practices from raw material extraction to customer feedback, business processes, and information exchange. It will showcase state-of-the-art technological advancements for enhancement of product quality, process productivity, and sustainability. The book series will include industry 4.0 and sustainability interventions in industrial engineering, systems, and management area. The series is aimed at senior undergraduate students, graduate students, academic researchers, and professionals in diverse areas including mechanical engineering, industrial engineering, human factors, ergonomics, and manufacturing engineering.

Supply Chain Management
Strategic Implementation in Manufacturing
Bharti, Shikha Singh, Anand Pandey and Amit Sachan

Industry 4.0 for Manufacturing Systems
Concepts, Technologies, and Applications
Vishal Ashok Wankhede, Vimal K.E.K and Pankaj Sahlot

Industry 4.0 for Manufacturing Systems

Concepts, Technologies, and Applications

Edited by
Vishal Ashok Wankhede, Vimal K.E.K.,
and Pankaj Sahlot

CRC Press
Taylor & Francis Group
Boca Raton London New York

CRC Press is an imprint of the
Taylor & Francis Group, an **informa** business

Designed cover image: Medium.com

First edition published 2025
by CRC Press
2385 NW Executive Center Drive, Suite 320, Boca Raton FL 33431

and by CRC Press
4 Park Square, Milton Park, Abingdon, Oxon, OX14 4RN

CRC Press is an imprint of Taylor & Francis Group, LLC

First edition published by CRC Press 2025

ISBN: 9781032744520 (hbk)
ISBN: 9781032747828 (pbk)
ISBN: 9781003470861 (ebk)

DOI: 10.1201/9781003470861

Typeset in Sabon
by Newgen Publishing UK

Contents

About the editors ix
List of contributors xi

1 Industry 4.0: transforming communication, sustainability, and collaboration in manufacturing 1
SANJAY KAUSHAL
1.1 *Introduction to Industry 4.0 1*
1.2 *Historical context and evolution of Industry 4.0 2*
1.3 *Communication impact of Industry 4.0 tools and techniques 2*
1.4 *Design principles and benefits of Industry 4.0 3*
1.5 *Communication in advanced manufacturing systems – the role of human–machine interaction 5*
1.6 *Data-driven communication: managing information flows and knowledge sharing in Industry 4.0 6*
1.7 *Communicating sustainability in Industry 4.0: toward a greener future 8*
1.8 *Conclusion 9*

2 Integrating Industry 4.0 technologies in manufacturing systems 12
ROHIT AGRAWAL AND RASHMI NAIR
2.1 *Introduction 12*
2.2 *The role of digital technologies 12*
2.3 *Integration of Industry 4.0 and additive manufacturing 13*
2.4 *Integration of Industry 4.0 and circular economy 13*
2.5 *Integration of Industry 4.0 and manufacturing systems 14*
2.6 *Integration of Industry 4.0 and Lean Six Sigma 15*
2.7 *Integration to other paradigms 16*
2.8 *Benefits and opportunities 16*

2.9 *Barriers to adoption of Industry 4.0 in large and small or medium enterprises 16*
2.10 *Frameworks and strategies for implementing Industry 4.0 18*
2.11 *Discussion 19*
2.12 *Conclusion 21*

3 Smart manufacturing 23
PARANITHARAN K.P., RAJA CHANDRA SEKAR M., RAMESH BABU T., BALAJI V., THANUSHKAR S., AND SREE RAMNATH SUDALAI S.
3.1 *The smart manufacturing concept 23*
3.2 *Drivers for adoption of smart manufacturing in large and small or medium enterprises 25*
3.3 *Benefits of the adoption of smart manufacturing in large and small or medium enterprises 28*
3.4 *Impact of smart manufacturing on sustainable development 28*
3.5 *Transition to smart manufacturing 30*
3.6 *Intelligent automation system 33*
3.7 *Developing smart manufacturing in industries 36*
3.8 *Conclusion 36*

4 Advanced manufacturing systems and Industry 4.0 42
NATARAJAN MANIKANDAN AND PASUPULETI THEJASREE
4.1 *Introduction 42*
4.2 *Concept of Industry 4.0 45*
4.3 *Conclusions 49*

5 Cyber-physical system for advanced manufacturing 54
RASHMI NAIR, PRIYANKA B. GAIKWAD, AND ROHIT AGRAWAL
5.1 *Introduction 54*
5.2 *Principles and applications 55*
5.3 *CPS integration with Industry 4.0 technologies 55*
5.4 *Context-aware computing for CPS 57*
5.5 *Cyber-physical systems in industrial robotics 58*
5.6 *Green CPS 59*
5.7 *Barriers to adoption of CPS 60*
5.8 *Future trends of CPS in Industry 4.0 61*
5.9 *Discussion 62*
5.10 *Conclusion 62*

6 Digital twins for advanced manufacturing 65
PRIYA SUBHA M., SATHISH V., AND JANARDHAN VISTAPALLI
6.1 *Introduction 65*
6.2 *Review on digital twins for various aspects of manufacturing 68*
6.3 *Elements of a digital twin 69*
6.4 *Frameworks for developing a digital twin 72*
6.5 *Case study of developing digital twin for a manufacturing system 79*
6.6 *Conclusions and future scope 84*

7 Decision-making in Industry 4.0 89
THEOPHILUS DHYANKUMAR C., JOSEPH FRANCIS J., AND SIVAKUMAR K.
7.1 *Introduction 89*
7.2 *Literature review 90*
7.3 *Decision-making using Industry 4.0 93*
7.4 *Case studies of effective decision-making in Industry 4.0 100*
7.5 *Managerial implications of decision-making in Industry 4.0 101*
7.6 *Conclusion 102*

8 Navigating the legal landscape of sustainable Industry 4.0: challenges and considerations 109
SOUMYA PRAKASH PATRA
8.1 *Introduction 109*
8.2 *Background and context 110*
8.3 *Legal challenges in sustainable Industry 4.0 112*
8.4 *Case studies: navigating legal challenges in sustainable Industry 4.0 115*
8.5 *Future directions 117*
8.6 *Conclusion 118*

9 Industry 4.0 performance measurement using key performance indicators for effective digital transformation 122
NISHAL MURALI AND RAM PRASAD KRISHNAKUMAR
9.1 *Introduction 122*
9.2 *I4.0 technologies 123*
9.3 *Importance of key performance indicators 126*
9.4 *Integration of Industry 4.0 tools and KPIs 130*
9.5 *Conclusion and future work 132*

10 Machine learning applications in inventory management: a case study 137
PRAKASH KUMAR, VRISHABH NARENDRA BHONDE, SONU RAJAK, AND MALOLAN SUNDARARAMAN

10.1 Introduction 137
10.2 Literature review 139
10.3 Methodology 140
10.4 Results and discussion 152
10.5 Conclusion 157

11 Research issues in Industry 4.0 160
PRIYANKA B. GAIKWAD AND VISHAL ASHOK WANKHEDE

11.1 Introduction 160
11.2 Current research challenges 163
11.3 Sector-specific research questions and proposition 166
11.4 Future directions and research opportunities 169
11.5 Conclusion 171

Index 174

About the editors

Vishal Ashok Wankhede is Assistant Professor in the Department of Operations Management and Quantitative Techniques at the Indian Institute of Management, Bodh Gaya, India, since July 2023. Previously, he served as Assistant Professor in the Department of Mechanical Engineering at Pandit Deendayal Energy University (PDEU), Gandhinagar, Gujarat, from March 2022 to June 2023, and in the Department of Industrial Engineering at PDEU from July 2016 to June 2019. Dr. Wankhede earned his Ph.D. in Industry 4.0 from the Department of Production Engineering, National Institute of Technology (NIT), Tiruchirappalli, in July 2022. He also holds an M.Tech. in Cloud Manufacturing (2016) and a B.Tech. in Mechanical Engineering (2014). He was recognized with the "Budding Researcher Award" in 2022 and an academic proficiency prize for his M.Tech. at NIT Trichy. He has published over 50 peer-reviewed articles and book chapters with over 1300 citations and an h-index of 20. He is a member of several professional organizations, including the Institution of Engineers (India) and the American Society of Mechanical Engineers. His research interests include Industry 4.0, smart manufacturing, additive manufacturing, circular economy, advanced manufacturing processes, optimization, and multi-criteria decision-making.

Google Scholar: https://scholar.google.com/citations?user=mynZ-eYAAAAJ&hl=en

Vimal K.E.K. received a Ph.D. degree in Sustainable Manufacturing in 2016 and an M.Tech. degree in Industrial Engineering in 2012 from the National Institute of Technology Tiruchirappalli, India, and a Production Engineering degree in 2010 from the PSG College of Technology, India. He has more than 10 years of teaching and research experience in Mechanical Engineering with a special emphasis on circular supply chain management, Lean Six Sigma, and sustainability in manufacturing. He has worked as a guest editor for a few journals. Presently, he acts as a reviewer for more than 20 prestigious Web of Science journals. In addition, he has also published (as an author and a co-author) 4 books with publisher, 20 book chapters, and more than 60 articles in journals and conferences. He is currently working as Assistant Professor, Department of Production Engineering, National Institute of Technology Tiruchirappalli, India.

Google Scholar: https://scholar.google.com/citations?user=ge6a3KEAAAAJ&hl=en&oi=sra

Pankaj Sahlot is Assistant Professor in the Mechanical Engineering Department at the National Institute of Technology (NIT) Kurukshetra, India. He also worked as Assistant Professor at Pandit Deendayal Energy University (PDEU), Gandhinagar, for around 4 years. He completed his Ph.D. from IIT Gandhinagar, India. He also visited the University of North Texas Denton, USA, for 6 months as a visiting researcher during his Ph.D. His research work has been published in various reputable journals and conferences. He did his M.Tech. from IIT Hyderabad in the Department of Materials Science and Engineering and received the Academic Excellence Award from Prof. C. N. R. Rao. He completed his B.Tech. in Mechanical Engineering at Ideal Institute of Technology in Ghaziabad, Uttar Pradesh, India. Currently, his research of interest focuses on additive manufacturing (AM) and Friction Stir Welding (FSW) of different metals. He has also collaborated with reputed national and international institutes and published several articles.

Google Scholar: https://scholar.google.com/citations?user=dpHiWzUAAAAJ&hl=en&oi=ao

Contributors

Rohit Agrawal
Operations Management and Quantitative Techniques
Indian Institute of Management Bodh Gaya
Bihar, India

Vrishabh Narendra Bhonde
Department of Production Engineering
National Institute of Technology Tiruchirappalli
Tamil Nadu, India

Theophilus Dhyankumar C.
Centre for Logistics and Supply Chain Management
Loyola Institute of Business Administration
Loyola College Campus, Chennai
Tamil Nadu, India

Priyanka B. Gaikwad
Department of Mathematics
Phulsing Naik Mahavidyalaya, Pusad
Maharashtra, India

Joseph Francis J.
Centre for Logistics and Supply Chain Management
Loyola Institute of Business Administration
Loyola College Campus, Chennai
Tamil Nadu, India

Ram Prasad Krishnakumar
Department of Mechanical Engineering,
Sri Venkateswara College of Engineering,
Sriperumbudur, Tamilnadu, India

Sivakumar K.
Centre for Logistics and Supply Chain Management
Loyola Institute of Business Administration
Loyola College Campus, Chennai
Tamil Nadu, India

Paranitharan K.P.
Department of Quality Assurance
TVS Sensing Solutions Private Limited
Tamil Nadu, India

Sanjay Kaushal
Humanities and Liberal Arts
Indian Institute of Management Bodh Gaya
Bihar, India

Prakash Kumar
Department of Production Engineering
National Institute of Technology Tiruchirappalli
Tamil Nadu, India

Nishal Murali
Department of Mechanical Engineering
Sri Venkateswara College of Engineering
Sriperumbudur, Tamil Nadu, India

Priya Subha M.
Department of Mechanical and Aerospace Engineering
Mahindra University
Bahadurpally Jeedimetla, Telangana, India

Natarajan Manikandan
Department of Mechanical Engineering
School of Engineering
Mohan Babu University
Tirupati, Andhra Pradesh, India

Raja Chandra Sekar M.
Department of Mechanical Engineering
Velammal College of Engineering and Technology
Madurai, Tamil Nadu, India

Rashmi Nair
School of Computing
MIT-Art, Design and Technology, Pune
Maharashtra, India

Soumya Prakash Patra
Humanities and Liberal Arts
Indian Institute of Management Bodh Gaya
Bihar, India

Sonu Rajak
Department of Mechanical Engineering
National Institute of Technology Patna
Bihar, India

Sree Ramnath Sudalai S.
Department of Mechanical Engineering
Velammal College of Engineering and Technology
Madurai, Tamil Nadu, India

Thanushkar S.
Department of Power Engineering
Brandenburgische Technische Universität Cottbus-Senftenberg
Brandenburgische, Germany

Malolan Sundararaman
Department of Management Studies
National Institute of Technology Tiruchirappalli
Tamil Nadu, India

Ramesh Babu T.
Department of Industrial Engineering
Anna University
Chennai, Tamil Nadu, India

Pasupuleti Thejasree
Department of Mechanical Engineering
School of Engineering
Mohan Babu University
Tirupati, Andhra Pradesh, India

Balaji V.
Department of Corporate Cell
TVS Sensing Solutions Private Limited
Madurai, Tamil Nadu, India

Sathish V.
Department of Mechanical and Aerospace Engineering
Mahindra University
Bahadurpally Jeedimetla, Telangana, India

Janardhan Vistapalli
Department of Mechanical and Aerospace Engineering
Mahindra University
Bahadurpally Jeedimetla, Telangana, India

Vishal Ashok Wankhede
Operations Management and Quantitative Techniques
Indian Institute of Management Bodh Gaya
Bihar, India

Chapter 1

Industry 4.0

Transforming communication, sustainability, and collaboration in manufacturing

Sanjay Kaushal

1.1 INTRODUCTION TO INDUSTRY 4.0

Industry 4.0 represents the fourth revolution in manufacturing and industry, emerging from the integration of digital information technologies with traditional industrial processes. This new era is characterized by the widespread adoption of cyber-physical systems (CPSs), the Internet of Things (IoT), and big data analytics, which collectively enhance communication, monitoring, and management of industrial operations. One of the hallmark examples of Industry 4.0 in action is Siemens' digital factory in Amberg, Germany. At this facility, systems and machines are interconnected, with products and machines exchanging information, driving decisions, and triggering actions autonomously. The plant boasts a remarkable defect rate of just 12 per million, showcasing the potential of smart manufacturing systems to improve quality and efficiency. This revolution extends beyond manufacturing; it encompasses a range of sectors including logistics and supply chains, evidenced by Amazon's use of Kiva robots in its warehouses to optimize package handling and reduce delivery times.

The defining features of Industry 4.0 include interconnectivity, automation, machine learning, and real-time data. This evolution is significantly transforming industries by enabling more flexible, responsive, and interconnected enterprises capable of making more informed decisions. For example, General Electric's Predix platform exemplifies this shift by providing a cloud-based platform that collects and analyzes data from industrial machines to predict maintenance issues before they occur, thereby minimizing downtime and reducing costs. Another instance is Ford's use of connected robots in assembly lines that work alongside humans to enhance productivity and reduce the physical strain on human workers. There is now a dual focus of Industry 4.0 on both technological advancement and enhancing the role of human workers through collaboration with machines. This synergy is crucial as it leverages the strengths of both human intuition and robotic precision, leading to innovation and efficiency in modern industrial practices.

DOI: 10.1201/9781003470861-1

1.2 HISTORICAL CONTEXT AND EVOLUTION OF INDUSTRY 4.0

Evolution of Industry 4.0 can be traced back to the initial industrial revolutions that fundamentally transformed society. The first industrial revolution in the late 18th century introduced mechanization through water and steam power, revolutionizing textile production and establishing the foundation for industrial activity. The second revolution, fueled by electricity in the early 20th century, ushered in mass production with assembly line methods, epitomized by Henry Ford's automobile factories, which significantly boosted production efficiency and brought products to the masses. The third revolution began in the 1970s with the advent of computers and the beginning of automation in manufacturing processes. Robots and programmable logic controllers entered factories, increasing production rates and efficiency, as seen in the automotive industry where robotic arms performed welding and painting tasks previously done by humans.

Industry 4.0, often referred to as the fourth industrial revolution, builds on this technological lineage by integrating digital technology into every area of business, providing new ways of creating value and reshaping manufacturing, supply chains, and management. Unlike previous revolutions, which were characterized by advances in energy and production, Industry 4.0 emphasizes deep connectivity and interactions between physical and digital systems. An example of this integration is the use of digital twin technology by companies like Bosch, which creates virtual models of physical machines to simulate, predict, and optimize the machine performance before actual physical deployment. Additionally, Airbus has been using advanced robotics and data analytics in its aircraft production lines to tailor operations and increase productivity, demonstrating the shift from traditional manufacturing to a connected, flexible, and data-driven production environment. These advancements reflect the core of Industry 4.0, which leverages connectivity, artificial intelligence (Kaushal & Mishra, 2024), and real-time data to drive further industrial innovation, thereby creating smarter, more efficient manufacturing ecosystems.

1.3 COMMUNICATION IMPACT OF INDUSTRY 4.0 TOOLS AND TECHNIQUES

The impact of Industry 4.0 tools and techniques on communication, particularly in bridging the gap between technology and human interaction, is substantial. Industry 4.0 has introduced sophisticated communication technologies that enable seamless interaction within factories and across global supply chains. For example, the implementation of CPSs and the IoT has revolutionized how machines, systems, and people communicate. Furthermore, the integration of Internet of Services (IoS) has allowed for enhanced service offerings and improved customer interactions by leveraging data analytics and cloud services. This evolution in communication technologies nurtures a more collaborative and integrated working environment, where humans and machines coexist and cooperate effectively, driving productivity and innovation (Marcon et al., 2017; Hauer, Harte, & Kacemi, 2018).

Moreover, the communication impact of Industry 4.0 extends beyond operational enhancements to influence strategic business decisions and human resource practices. The integration of digital platforms and tools in communication has led to the development of new competencies in the workforce. For instance, employees today are required to possess a blend of technical skills and advanced communication capabilities to steer the complex sphere of digital manufacturing. Companies are increasingly investing in training and development programs to equip their staff with necessary digital skills, including cognitive analytics, data management, and technology literacy, all critical for managing the nuances of Industry 4.0 technologies. Such educational initiatives highlight the commitment to not only advancing technological proficiency but also enhancing communication skills, which are indispensable in the highly interconnected and digitized industrial setups of today (Lee & Meng, 2021). These efforts demonstrate the ongoing transformation in workplace dynamics, driven by Industry 4.0, where effective communication is the cornerstone for successful technology integration and human-machine collaboration.

1.4 DESIGN PRINCIPLES AND BENEFITS OF INDUSTRY 4.0

Industry 4.0 represents a significant leap forward in the connectivity of all processes and products across their entire lifecycle, fundamentally redefining production dynamics. In this ecosystem, human operators, machines, products, and processes all interact in a self-organizing, interconnected manner. This connectivity facilitates an integrated response to changes and enhances ongoing planning within management systems. Viewed as a manufacturing revolution, Industry 4.0 aims to elevate productivity and increase stakeholder value. However, despite its potential, it is often misunderstood as a standalone solution to existing problems. Rather, Industry 4.0 demands a paradigm shift within companies to foster innovation, maintain competitiveness, and boost productivity. To achieve the required autonomy for this revolution, guided by frameworks like the Reference Architecture Model for Industry (RAMI 4.0) and the Intelligent Manufacturing Systems Architecture (IMSA), we face challenges such as a general lack of understanding and acceptance of standards, as evidenced by minimal citations of the RAMI 4.0 standards. Nonetheless, RAMI 4.0 offers a comprehensive map for horizontal and vertical integration across the product lifecycle, ensuring secure network data exchange. Such integration increases transparency and communication across the entire value chain, suggesting new business models centered around value addition for all stakeholders. However, implementing Industry 4.0 poses financial risks, particularly for small- and medium-sized enterprises (SMEs) deterred by the high costs associated with advanced technology and IT infrastructure (Davis et al., 2020). Central to Industry 4.0 are following six key design principles; these principles are foundational for advancing the capabilities and effectiveness of Industry 4.0 systems:

1.4.1 **Interoperability** enables seamless interaction between connected products and systems, allowing machines to exchange and interpret data without requiring additional effort from users, thus enhancing integration across services, products, and processes.
1.4.2 **Virtualization** involves creating digital replicas of physical systems, enabling end-to-end visibility across the product lifecycle, which aids in optimizing production planning and identifying potential errors early in the design phase.
1.4.3 **Decentralization** allows decision-making to be pushed to lower levels of the production process, fostering a responsive and adaptive system that can self-organize and react promptly to changes, enhancing productivity and information flow.
1.4.4 **Real-time capability** is crucial for monitoring and processing data instantaneously, which is vital for both decentralized decision-making and effective virtualization, helping to reduce reaction times and enhance productivity.
1.4.5 **Service orientation** facilitates access to information and real-time adjustments based on customer preferences, increasing transparency in the value chain and focusing on customer-oriented service delivery.
1.4.6 **Modularity** promotes standardization across different systems, enabling compatibility and flexibility in production, which supports mass customization and rapid response to changes.

The adoption of Industry 4.0 offers a multitude of benefits that contribute to the advancement of manufacturing systems and processes. Increased productivity is a notable advantage, as advanced automation and optimized processes facilitate higher production rates and more efficient use of resources. Industry 4.0 also promotes improved quality control through precise monitoring and data analytics, resulting in a reduction of defects and rework. In a practical application, several companies utilize data analytics to ensure consistent quality across their manufacturing lines. Cost savings are another key benefit, with efficient resource allocation and predictive maintenance reducing downtime and overall operational expenses. General Electric (GE), for example, uses digital twin technology to optimize machine performance and predict maintenance needs, leading to lower costs associated with equipment failure. The paradigm also enhances flexibility and customization, allowing manufacturers to rapidly adjust production lines to meet specific customer demands. Moreover, Industry 4.0 furthers better decision-making through access to real-time data and actionable insights, enabling managers to make informed strategic choices. Sustainability is also a prominent benefit, as efficient resource management and smart energy usage contribute to a reduction in waste and environmental impact.

Improved workplace safety is achieved through advanced technologies such as robotics and IoT, which minimize human exposure to hazardous tasks. For example, Amazon's use of robots in its fulfillment centers enhances safety and productivity; Bosch's implementation of IoT platforms that facilitate communication between machines and operators ensures that critical data regarding system performance and potential faults is instantly available, thus supporting faster and more informed decision-making across the organization (Habib & Chimsom, 2019; Cañas et al., 2021). Industry 4.0 also encourages the emergence of innovative business models, shifting toward data-driven, service-oriented approaches that open up new revenue streams and growth opportunities. This combination of benefits positions Industry 4.0 as a transformative force in modern manufacturing, aligning with the needs of an increasingly dynamic and competitive marketplace.

1.5 COMMUNICATION IN ADVANCED MANUFACTURING SYSTEMS – THE ROLE OF HUMAN-MACHINE INTERACTION

In contemporary manufacturing ecosystems, the importance of human-machine interaction (HMI) is central to achieving operational excellence, ensuring safety, and enhancing flexibility. Sophisticated HMI systems equip human operators with the means to engage seamlessly with complex machinery and automated systems, facilitating a higher degree of precision and control in manufacturing processes. For instance, in the context of smart factories, operators utilize advanced interactive dashboards that relay crucial real-time data concerning machine performance, production statistics, and potential system errors. This level of interaction permits immediate human responses to fine-tune operations or rectify issues, effectively minimizing downtime and sustaining productivity levels. A study by Joo and Shin (2019) highlights the intricacies of HMIs within smart manufacturing settings, proposing that adaptive automation can refine the interplay between human operators and machines. By enhancing clarity in communication and boosting the effectiveness of manufacturing oversight, adaptive automation significantly elevates the cooperative dynamics within production systems, marrying human expertise with machine efficiency to foster superior production outcomes.

Further elaboration on the design of HMI systems reveals a strong emphasis on ergonomic principles to ensure that interactions between human operators and machines are both comfortable and safe, particularly during extended operations. Ergonomic HMIs are instrumental in alleviating operator fatigue and physical strain, which in turn boosts job satisfaction and overall productivity. One illustrative advancement in this area is the application of virtual reality (VR) and augmented reality (AR) technologies, which are employed to recreate manufacturing scenarios for training purposes. Through these technologies, operators are able to rehearse intricate procedures within a controlled virtual environment, mitigating the risk of equipment damage or safety breaches. Research by Cochran et al. (2017) highlights the significance of incorporating HMI considerations into equipment design, not only to enhance the physical workspace by aligning it with human physiological capabilities but also to integrate the behavioral roles of operators. This integration boosts operator engagement and amplifies their effectiveness in roles that require supervisory control. These developments in HMI design signify a significant shift toward more interactive and dynamic manufacturing environments, where human cognitive and physical strengths are complementarily augmented by machine intelligence, paving the way for smarter, safer, and more adaptive production systems.

1.6 DATA-DRIVEN COMMUNICATION: MANAGING INFORMATION FLOWS AND KNOWLEDGE SHARING IN INDUSTRY 4.0

In the context of Industry 4.0, data-driven communication is essential for managing information flows and enhancing knowledge sharing across manufacturing enterprises. This modern industrial framework is characterized by the integration of technologies such as the IoT, big data analytics, and cloud computing, which collectively streamline the capture, analysis, and dissemination of vast amounts of data.

For instance, Meski et al. (2019) discuss the implementation of new information and communication technologies (ICTs) in manufacturing floors, which generate substantial data, necessitating robust data and knowledge management approaches. These systems facilitate real-time decision-making and enhance operational efficiency by enabling seamless communication between machines and human operators (Kaushal et al., 2023), thereby optimizing the entire production chain from inventory management to quality control. Leading manufacturers like Siemens, ABB, Rockwell Automation, Schneider Electric, and many others have successfully implemented such ICT-driven communication systems, integrating their operations, supply chain, and maintenance functions to achieve greater efficiency and responsiveness.

Moreover, the role of collaborative data analytics (CDAs) in Industry 4.0 is transformative, as highlighted by Lazarova-Molnar et al. (2019). They propose a CDA framework that supports manufacturing enterprises of all sizes in sharing and analyzing data collectively to improve decision-making processes. This collaboration is particularly beneficial for SMEs that may lack the extensive data resources of larger companies. By pooling their data, SMEs can achieve a more comprehensive analysis, leading to better operational insights and competitive advantages. Teixeira et al. (2018) emphasize the integration of Lean Thinking in managing information flows within Industry 4.0, proposing a framework that aligns with efficient data handling to minimize waste and maximize value creation. This methodology ensures that information flows are not just fast and continuous but also strategically aligned with the organizational goals, enhancing overall productivity and reducing inefficiencies. Their approach reiterates the importance of structured information management in achieving operational excellence and sustainable business practices in the digital era. Manufacturers like Toyota and Danaher have successfully combined Lean principles with Industry 4.0 technologies to streamline their data-driven communication and decision-making processes.

In addition to structured data management, the incorporation of big data into knowledge management is crucial for organizations adapting to Industry 4.0, as discussed by Cárdenas et al. (2018). They propose a model that utilizes big data tools to manage the vast information flows characteristic of Industry 4.0, helping organizations gain competitive and comparative advantages. This model highlights the critical nature of technological solutions in analyzing extensive data sets, which is key for developing strategic initiatives and maintaining a competitive edge in a rapidly evolving market. Companies like GE Digital and PTC have developed industry-leading big data analytics platforms that enable their manufacturing clients to derive actionable insights from the vast amounts of data generated across their operations. Pedro et al. (2022) introduce a novel information sharing system using linked data, ontologies, and knowledge graph technologies to improve safety outcomes in the construction industry. Their system exemplifies how advanced data integration and semantic modeling can enhance the accessibility and utility of information, facilitating better knowledge sharing and learning across sectors. This illustrates the broad applicability of data-driven communication strategies in Industry 4.0, not only in manufacturing but also in other industries where safety and efficiency are paramount. Innovative players like Bentley Systems and Autodesk have pioneered the use of knowledge graph and semantic technologies to enhance information sharing and collaboration in the construction and infrastructure sectors. These recent developments in data-driven communication within Industry 4.0 highlight the potential of integrating advanced digital technologies to manage information flows and enhance organizational knowledge sharing (Kaushal, Nyoni, & Sharma, 2024).

1.7 COMMUNICATING SUSTAINABILITY IN INDUSTRY 4.0: TOWARD A GREENER FUTURE

In the age of Industry 4.0, the conversation around sustainability has intensified, as this era offers unprecedented opportunities to integrate eco-friendly practices with advanced technological processes. Industry 4.0, characterized by automation and data exchange in manufacturing technologies, includes CPSs, the IoT, cloud computing, and cognitive computing. These technologies facilitate more efficient resource usage and energy consumption, contributing significantly to environmental sustainability. For example, predictive maintenance enabled by IoT devices can dramatically reduce downtime and save energy by servicing machinery only when necessary, rather than on a fixed schedule regardless of actual need. This approach not only enhances operational efficiency but also minimizes the environmental footprint of manufacturing activities. Research by Tiwari and Khan (2020) explores how sustainability accounting and reporting in Industry 4.0 can be leveraged to enhance ecological and corporate responsibility, pointing toward a strategic integration of sustainability goals within industrial processes. The implementation of sustainable practices in Industry 4.0 also emphasizes the economic benefits alongside environmental protection. For instance, the optimization of supply chains through real-time data analytics and automation technologies can lead to significant reductions in waste and energy consumption, thereby supporting the triple bottom line of people, profit, and the planet. Yildiz Çankaya and Sezen (2020) discuss how modern industry developments, spurred by Industry 4.0 technologies, necessitate new business models that integrate sustainability into core business strategies. These models are crucial for ensuring long-term environmental, economic, and social value creation, illustrating a shift from traditional to sustainable business practices driven by technological advancements. Companies like Dassault Systèmes have successfully implemented Industry 4.0-enabled supply chain optimization solutions, leading to improved sustainability metrics and increased profitability.

Furthermore, real-time monitoring and control systems provided by Industry 4.0 technologies allow for immediate responses to operational or environmental changes, thus promoting sustainability. For example, Ghobakhloo (2020) identifies how digital transformation within Industry 4.0 frameworks supports sustainability by improving production efficiency and fostering innovations that reduce harmful environmental impacts. This research highlights the role of Industry 4.0 in achieving sustainability functions that align with the United Nations Sustainable Development Goals (SDGs). Companies like Emerson Process Management and Honeywell Process Solutions have developed comprehensive Industry 4.0 solutions that integrate real-time monitoring, control, and communication capabilities, enabling their manufacturing clients to quickly identify and address sustainability-related issues. Moreover, Industry 4.0 offers a unique platform for advancing sustainable manufacturing through the development and application of smart technologies. Habib and Chimsom (2019) explore the potential of these technologies to increase system intelligence, agility, and flexibility, which are key for sustainable development. By enhancing the interoperability and modular nature of manufacturing systems, Industry 4.0 technologies can significantly contribute to sustainable industrial practices. This transition not only supports environmental goals but also promotes social sustainability by creating safer and more engaging work environments.

As industries embrace the fourth industrial revolution, the integration of sustainability in Industry 4.0 becomes imperative for maintaining competitiveness and securing a sustainable future. The challenge lies in effectively communicating and implementing these sustainability strategies within the framework of Industry 4.0 to ensure they are deeply embedded in all levels of industrial operations (Nyoni & Kaushal, 2022a). Bai et al. (2020) provide a comprehensive assessment of Industry 4.0 technologies from a sustainability perspective, suggesting that while these technologies hold great promise for sustainable development, each must be carefully evaluated to understand its specific impacts and benefits. This careful evaluation will ensure that the transition to smarter manufacturing will not only be technologically advanced but also sustainably responsible, thereby contributing to a greener future.

1.8 CONCLUSION

Industry 4.0 represents a transformative leap in the manufacturing sector, fundamentally reshaping communication, collaboration, and sustainability. By integrating advanced technologies such as CPSs, IoT, and big data analytics, Industry 4.0 enhances real-time communication and decision-making, driving operational excellence and nurturing innovation. This revolution is not just technological but also deeply human-centric, emphasizing the synergy between human operators and intelligent machines. Moreover, Industry 4.0's commitment to sustainability is evident in its potential to optimize resource use, reduce environmental impact, and create new business models aligned with the triple bottom line. However, realizing the full potential of Industry 4.0 requires a holistic approach that integrates these technologies with sustainable practices and addresses the challenges of implementation, particularly for SMEs. As industries continue to evolve, the success of Industry 4.0 will hinge on its ability to balance technological advancement with sustainable and ethical practices, ensuring a greener and more efficient future for manufacturing.

REFERENCES

Bai, C., Dallasega, P., Orzes, G., Sarkis, J., & Sarkis, J. (2020). Industry 4.0 technologies assessment: A sustainability perspective. *International Journal of Production Economics*, 229, 107776. https://doi.org/10.1016/j.ijpe.2020.107776

Cañas, H., Mula, J., Díaz-Madroñero, M., & Campuzano, F. (2021). Implementing Industry 4.0 principles. *Computers & Industrial Engineering*, 158, 107379. https://doi.org/10.1016/J.CIE.2021.107379.

Cárdenas, L.J.A., Ramírez, W.F.T., & Rodríguez Molano, J.I. (2018). Model for the incorporation of big data in knowledge management oriented to Industry 4.0. In Y. Tan, Y. Shi, & Q. Tang (Eds.), *Data Mining and Big Data. DMBD 2018* (pp. 683–693) *Lecture Notes in Computer Science*, vol 10943. Springer, Cham. https://doi.org/10.1007/978-3-319-93803-5_64

Cochran, D., Arinez, J., Collins, M., & Bi, Z. (2017). Modelling of human–machine interaction in equipment design of manufacturing cells. *Enterprise Information Systems*, 11, 969–987. https://doi.org/10.1080/17517575.2016.1248495

Davis, N., Companiwala, A., Muschard, B., & Petrusch, N. (2020). 4th Industrial revolution design through Lean foundation. *Procedia CIRP*, 91, 306–311. https://doi.org/10.1016/j.procir.2020.03.102

Ghobakhloo, M. (2020). Industry 4.0, digitization, and opportunities for sustainability. *Journal of Cleaner Production*, 252, 119869. https://doi.org/10.1016/j.jclepro.2019.119869

Habib, M., & Chimsom, C. (2019). Industry 4.0: Sustainability and design principles. *2019 20th International Conference on Research and Education in Mechatronics (REM)*, 1–8. https://doi.org/10.1109/REM.2019.8744120

Hauer, G., Harte, P., & Kacemi, J. (2018). An exploration of the impact of Industry 4.0 approach on corporate communication in the German manufacturing industry. *International Journal of Supply Chain Management*, 7(4), 125–131.

Joo, T., & Shin, D. (2019). Formalizing human–machine interactions for adaptive automation in smart manufacturing. *IEEE Transactions on Human-Machine Systems*, 49, 529–539. https://doi.org/10.1109/THMS.2019.2903402

Kaushal, S., & Mishra, D. (2024). Strategic implications of AI in contemporary business and society. *Proceedings of the 2nd International Conference on Disruptive Technologies (ICDT)*, Greater Noida, India, 1469–1474. https://doi.org/10.1109/ICDT61202.2024.10489027

Kaushal, S., Nyoni, A. M., & Sharma, A. (2024). Are we making progress in developing knowledge management strategies that support organizational performance? *Kybernetes. The International Journal of Cybernetics, Systems and Management Sciences, 53(12)*, 6097–6113. https://doi.org/10.1108/k-05-2023-0739

Kaushal, S., Nyoni, A. M., Sharma, A., & Saidi, J. A. K. (2023). Drawbacks of technology as an antecedent for knowledge sharing: Focus on curbing misinformation. *Proceedings of the 9th International Conference on Advanced Computing and Communication Systems (ICACCS)*, Coimbatore, India, 2483–2487. https://doi.org/10.1109/ICACCS57279.2023.10112895

Lazarova-Molnar, S., Mohamed, N., & Al-Jaroodi, J. (2019). Data analytics framework for Industry 4.0: Enabling collaboration for added benefits. *IET*

Collaborative Intelligent Manufacturing, 1(3), 117–125. https://doi.org/10.1049/iet-cim.2019.0012

Lee, J. J., & Meng, J. (2021). Digital competencies in communication management: A conceptual framework of Readiness for Industry 4.0 for communication professionals in the workplace. *Journal of Communication Management*, 25(4), 417–436. https://doi.org/10.1108/jcom-10-2020-0116

Marcon, P., Zezulka, F., Veselý, I., Szabó, Z., Roubal, Z., Sajdl, O., Gescheidtová, E., & Dohnal, P. (2017). Communication technology for Industry 4.0. *2017 Progress in Electromagnetics Research Symposium – Spring (PIERS)*, St. Petersburg, Russia, 1694–1697. https:/doi.org/10.1109/PIERS.2017.8262021

Meski, O., Belkadi, F., Laroche, F., Ladj, A., & Furet, B. (2019). Integrated data and knowledge management as key factor for Industry 4.0. *IEEE Engineering Management Review,* 47, 94–100. https://doi.org/10.1109/EMR.2019.2948589

Nyoni, A. M., & Kaushal, S. (2022a). Sustainable knowledge management during crisis: Focus on Covid-19 pandemic. *Business Information Review*, 39(4), 136–146. https://doi.org/10.1177/02663821221109928

Pedro, A., Pham-Hang, A., Nguyen, P., & Pham, H. (2022). Data-driven construction safety information sharing system based on linked data, ontologies, and knowledge graph technologies. *International Journal of Environmental Research and Public Health*, 19(2), 794. https://doi.org/10.3390/ijerph19020794

Teixeira, L., Ferreira, C., & Santos, B.S. (2018). An information management framework to Industry 4.0: A Lean Thinking approach, In T. Ahram, W. Karwowski, R. Taiar (Eds.), *Human Systems Engineering and Design. IHSED 2018. Advances in Intelligent Systems and Computing* (pp. 1063–1069, vol 876). Springer, Cham. https://doi.org/10.1007/978-3-030-02053-8_162

Tiwari, K., & Khan, M.S. (2020). Sustainability accounting and reporting in the Industry 4.0. *Journal of Cleaner Production*, 258, 120783. https://doi.org/10.1016/j.jclepro.2020.120783

Yildiz Çankaya, S. & Sezen, B. (2020). Industry 4.0 and Sustainability. In U. Akkucuk (Ed.), *Handbook of Research on Creating Sustainable Value in the Global Economy* (pp. 67–84). IGI Global Scientific Publishing. https://doi.org/10.4018/978-1-7998-1196-1.ch005

Chapter 2

Integrating Industry 4.0 technologies in manufacturing systems

Rohit Agrawal and Rashmi Nair

2.1 INTRODUCTION

Industry 4.0 refers to so-called fourth-generation operations that link manufacturing control systems and data with the "smart factories" of today, uniting digital technologies like the Internet of Things (IoT), artificial intelligence (AI), big data and analytics, and cyber-physical systems (CPSs). This new industrial model is based on the development of smart factories that are defined by interconnected, real-time data exchange and autonomous decision-making, which aims to improve productivity, efficiency, and flexibility (Kagermann et al., 2013).

It is the fourth industrial revolution after mechanization with water and steam, mass production with electricity, and automatic production with IT (Information Technology). Using advanced digital technologies, IoT enables the virtual and physical worlds to unify, creating a coordinated, fully connected system of the value chain, from product development to production to logistics, customer service, etc. (Hermann et al., 2016).

2.2 THE ROLE OF DIGITAL TECHNOLOGIES

Based on the existing literature, Industry 4.0 core technologies are IoT, AI, big data analytics, CPSs, and advanced robotics. Through IoT, machines, devices, and systems can connect to each other and communicate, thereby enabling the collection of an enormous amount of data and its real-time analysis. The data generated from these devices is then processed by the AI and machine learning algorithms to help in operation optimization and maintenance predictions for making better decisions (Xu et al., 2018). CPSs embed computational parts into physical processes and allow for real-time monitoring and control, while advanced robotics extend automation as well as increase cooperation of man and machinery (Monostori et al., 2016).

DOI: 10.1201/9781003470861-2

2.3 INTEGRATION OF INDUSTRY 4.0 AND ADDITIVE MANUFACTURING

2.3.1 Production flexibility improvement

One of the unique capabilities of additive manufacturing (AM) is the ability to create parts with complex geometries and to customize parts with little lead time. By connecting Industry 4.0 technologies with AM, better production flexibility can be achieved to provide real-time monitoring, data analytics, as well as adaptive manufacturing processes (Tao et al., 2018).

2.3.2 Digital twins and simulations

When integrated with AM, digital twins and simulation offer a virtual image of the manufacturing process. This integration permits a more streamlined design, testing, and production process and ultimately enables a reduction in time-to-market and an increase in product quality (Negri et al., 2017).

2.3.3 Data-driven decision-making

Advanced capabilities of Industry 4.0 technologies allow collecting and studying large amounts of data resulting from the AM process. Such a data-driven process is critical for predictive maintenance, quality assurance, and process optimization, all of which contribute to process efficiency and operational cost reductions (Tao et al., 2018).

2.3.4 Fabrication on demand

Industry 4.0 with AM creates a capability for mass customization with the idea of being able to produce small lots of highly individualized products for a low price. This is particularly important in sectors like health, aerospace, or automotive, encouraging the search for tailor-made solutions (Negri et al., 2017).

2.4 INTEGRATION OF INDUSTRY 4.0 AND CIRCULAR ECONOMY

2.4.1 Sustainable manufacturing practices

This drives sustainable manufacturing practices as it combines Industry 4.0 with circular economy principles. IoT, AI, and blockchain technologies track and trace materials in the lifecycle approach, supporting the recycling, remanufacturing, and reduction of waste (Pagoropoulos et al., 2017).

2.4.2 Resource efficiency

What Industry 4.0 technologies do best is that they improve resource efficiency, i.e., making the production process as efficient as possible, in terms of energy, a minimum amount of materials used, or waste generated. Precise predictions, pattern recognition, and the identification of inefficiencies prevent petrification, facilitating more sustainable manufacturing operations (Antikainen et al., 2018).

2.4.3 Product lifecycle management

The integration supports effective Product Life cycle Management, providing visibility to product usage, performance, and end-of-life options. This data can also be useful for manufacturers to design products with durable life, repairability, and recyclability aimed at fulfilling the principles of the circular economy (Antikainen et al., 2018).

2.4.4 Reverse supply chains

Even supply chains are moving towards closed-loop systems where materials and products are reused and recycled continuously with Industry 4.0. In an intelligent materials system, for instance, such as one supported by IoT and blockchain technologies, transparency and traceability can help to guarantee background about whether or not materials are truly intermeshed with once more within the production method (Pagoropoulos et al., 2017).

2.5 INTEGRATION OF INDUSTRY 4.0 AND MANUFACTURING SYSTEMS

2.5.1 Smart manufacturing systems

The companies or factories that used to have traditional manufacturing processes are now being treated as smart manufacturing systems through Industry 4.0, where they are communicating with each other in the era of IoT, AI, robotics, and advance big data analytics. The systems are self-optimizing, self-diagnosing, and able to modify their operations independently to increase efficiency and productivity (Kagermann et al., 2013).

2.5.2 Real-time monitoring and controlling

IOT sensors and devices allow real-time monitoring and control over the manufacturing processes. This integration delivers real-time information about production performance, equipment health, and process deviations, enabling better front-line decision-making (Kagermann et al., 2013).

2.5.3 Cyber-physical systems

The use of CPSs blurs the boundaries of the physical and digital world and allows systems, machines, and even humans to communicate effortlessly. CPSs improve the automation of processes and also decrease human error, which in turn increase overall manufacturing efficiency (Monostori et al., 2016).

2.5.4 Collaborative robotics

These are collectively called as collaborative robots (cobots), which can share the workspace with human operators, and their roles can be on activities that require precision, strength, and endurance. Integrating cobot technologies with Industry 4.0 ensures collision-free and effective collaboration, which helps the acceleration in increasing the productivity and lowering the risk of the worker injuries (Villani et al., 2018).

2.6 INTEGRATION OF INDUSTRY 4.0 AND LEAN SIX SIGMA

2.6.1 Amalgamation of Lean principles and digital technologies

Lean principles are upgraded with Industry 4.0 through digital tools for continuous improvement, waste reduction, and process optimization. By integrating IoT, data analytics, and AI, it will be possible to monitor and analyze the same production processes in real time to detect waste and bottlenecks (Sanders et al. 2016).

2.6.2 Data-driven Lean Six Sigma

The convergence of Industry 4.0 with Lean Six Sigma intersects big data and analytics to reach data-powered strategies. The methodology improves Lean Six Sigma methodologies, giving a more accurate and effective way to measure quality and process efficiency (Sony & Naik, 2019).

2.6.3 Predictive and prescriptive analytics

How predictive and prescriptive analytics help to blend Industry 4.0 with Lean Six Sigma? This type of advanced analytics can predict potential problems and process deviations, where prescriptive analytics can make courses of action recommendations for process improvements (Sanders et al., 2016).

2.6.4 Real-time process management and optimization

This is one of the critical Industry 4.0 technologies that supports real-time process control; it helps in quick corrective actions and ongoing process improvement. This capability is consistent with the Lean Six Sigma philosophy of reducing variation.

2.7 INTEGRATION TO OTHER PARADIGMS

Industry 4.0 is not independent but interacts synergistically with other industrial paradigms, including Lean Manufacturing, AM, and the Circular Economy. Combining lean principles with the advanced automation technology of Industry 4.0, for example, can improve efficiency and reduce waste through real-time data mining and process optimization (Sanders et al., 2016). By combining these technologies in the context of a circular economy and Industry 4.0, the possibilities of flexible, on-demand production and environmentally sustainable, resource-efficient manufacturing are improved (Pagoropoulos et al., 2017).

2.8 BENEFITS AND OPPORTUNITIES

Industry 4.0 offers several advantages such as streamline operation, reduced cost, product quality, and flexibility of the production process. Predictive maintenance can help manufacturers use real-time data to reduce downtime and increase their equipment lifespan. In addition, Industry 4.0 enables mass customization, a process where suppliers can cater to individual customer needs while maintaining efficiency (Lasi et al., 2014).

Additionally, Industry 4.0 supports sustainability through efficient utilization of available resources, minimization of waste, and lesser consumption of energy. Designing for reuse, remanufacturing, and recycling is part of the circular economy principles, which can be introduced to the Industry 4.0 paradigm to allow for proper closed-loop supply chains to be carried out, which decreases the overall environmental impact (Antikainen et al., 2018).

2.9 BARRIERS TO ADOPTION OF INDUSTRY 4.0 IN LARGE AND SMALL OR MEDIUM ENTERPRISES

The transition toward Industry 4.0 comes with some issues even though it brings a number of advantages. These obstacles are very real and include technical barriers like the difficulty of integrating new technologies with legacy infrastructure as well as the lack of interoperability among different systems. However, the financial limitations, especially for small- and medium-sized enterprises (SMEs), can restrict the investment in the technologies that are required and training (Mittal et al., 2018).

The implementation process is further complicated by organizational and cultural obstacles, such as a resistance to change, a workforce that lacks the digital skills and knowledge of new technologies, as well as a limited feeling for digitalization. Regulatory and compliance difficulties further complicate the situation, and, combined with data protection issues, require a solution strategy by a thorough approach (Moeuf et al., 2018).

2.9.1 Technological barriers

One of the biggest obstacles to incumbent enterprises and SMEs adopting the Industry 4.0 is this technological barrier. These are relatively things like the lack of interoperability among the technologies and systems, the limitations with the ICT infrastructure, and the complexity of interfacing with the legacy systems protecting sensitive data. This problem is exacerbated in the case of SMEs by the limited access to state of-the-art technologies and know-how (Schumacher et al., 2016).

2.9.2 Financial constraints

The adoption of Industry 4.0 depends on investing a lot of money in new technology, training, and changing the processes. Investing in these may prove a little harder for SMEs compared to large MNCs, which have more resources to allocate into such investments. Moreover, "many SMEs did not invest in and adopted Industry 4.0 (low level of agility and small scale of investment) due to the risk of investment in innovations without a real guarantee of returns" (Mittal et al., 2018).

2.9.3 Barriers at the organizational and cultural level

The adoption of Industry 4.0 requires profound changes in organizational culture and processes. Resistance from employees to change, a lack of digital skills, and no strategic vision from which to take inspiration can all obstruct tactics being put into practice. Sony and Naik (2019) indicated the risk of bureaucratic inertia for large enterprises and of lacking managerial capability to drive such transformations effectively for SMEs (Sony & Naik, 2019).

2.9.4 Regulatory and compliance concerns

Especially in the case of regulatory compliance and industry standards, this might get a bit tricky for Industry 4.0 technologies adoption. Such regulations will differ from country to country and among specific industry sectors, resulting in complicating more for multinationals. In particular, small and medium-sized entities (SMEs) may struggle to understand and comply with these regulations as they have limited resources and expertise available (Moeuf et al., 2018).

2.9.5 Security and privacy issues of data

Broadening the scope of this connectivity and data exchange also raises considerable questions about security and privacy in the context of the connected factory, which serves as a paradigm case of Industry 4.0. Even large enterprises with security measures in place risk data breaches and cyberattacks. These threats are particularly great for SMEs, those firms with their narrow resources (Janssen et al., 2020).

2.10 FRAMEWORKS AND STRATEGIES FOR IMPLEMENTING INDUSTRY 4.0

For any organization to enable Industry 4.0, they have to come up with necessary frameworks and strategies to consume these challenges. This includes understanding whether the company is ready for digital transformation, connecting digital initiatives with business objectives, and promoting innovation and continuous improvement. In order to drive digital transformation, it is important that the workforce becomes ready for the challenge and is able to adopt the new technology (Schumacher et al., 2016).

Engaging with outside partnerships, such as technology vendors, research organizations, and industry consortia, could also offer the necessary support and knowledge to help traverse the challenging adoption landscape for businesses entering the realm of Industry 4.0. Policymakers contribute to the creation of favorable regulatory surroundings and incentives so that you can improve the brand-new virtual infrastructure (Janssen et al., 2020).

A holistic framework of readiness assessment should include multiple dimensions – technology, organization, strategy, workforce, and external environment. The subdimensions assess characteristics of an enterprise that indicate its readiness for Industry 4.0 (Schumacher et al., 2016). Recent works on the conceptual framework to classify and measure Industry 4.0 readiness pertaining to advanced manufacturing are explained in the below sub-section.

2.10.1 Technology readiness

This dimension assesses the level of technology base of the actual infrastructure in terms of the availability of advanced manufacturing technologies, ICT systems, and data management capabilities. It measures the ability of an enterprise to build new technologies and integrate them with the existing systems (Mittal et al., 2018).

2.10.2 Preparedness of the organization

Organizational readiness is evaluating the internal processes, systems, and culture that support and align with driving digital transformation. The leadership commitment, change management capabilities, and having a digitally enabled strategy are few more important factors identified as the enablers (Sony & Naik, 2019).

2.10.3 Strategic readiness

Strategic readiness is the extent to which Industry 4.0 initiatives have their activities aligned in terms of overall Business Strategy. Are strategic objectives of the digital transformation process well defined? Is the road-mapping done properly? Are the resources committed to the digital transformation processes (Schumacher et al., 2016)?

2.10.4 Workforce readiness

One of the aspects that has the lowest workforce readiness rates is with respect to jobs requiring different new (and not-so-new) skills and competencies needed to work with Industry 4.0 technologies. For the HRM dimension, the indicator would be about the availability of training programs, the number of its employees participating in digital initiatives, and the capability to hire and retain talents with digital acumen (Mittal et al., 2018).

2.10.5 External environment readiness

Industry 4.0 adoption is examined based upon external pressures, including the regulatory, standards, and external support and partnership. It also takes into account the competitive landscape and market dynamics (Moeuf et al., 2018).

2.11 DISCUSSION

The Industrial Internet of Things (IIoT) and the Industry 4.0 technologies have revolutionized the manufacturing sector, leading to various opportunities and challenges in the implementation and integration of these technologies. This article discusses the impact of Industry 4.0 and reviews technological advancements, cultural change, economic impact, and future expectations.

2.11.1 Technological development and integration

Industry 4.0 utilizes a set of emerging technologies like IoT, AI, Big Data Analytics, and CPSs and also includes intelligent robotics to build Smart

Factories (Kagermann et al., 2013). These technologies enable dynamic information exchange, automation, and decision-making, which drives greater efficiency and productivity (Xu et al., 2018). Enabling an end-to-end and integrated IoT ecosystem, machines and systems seamlessly communicate addressing predictive maintenance and reduce downtime (Zhou et al., 2015).

2.11.2 Organizational change and workforce development

Industry 4.0, if implemented successfully, requires a major shift in the way the organizations operate. Organizations need to establish a coherent digital strategy, create an innovation-enabling environment, and invest in workforce development for enhancing the digital skills of their workforce (Schumacher et al., 2016). However, during the same period, resistance to change and the absence of digital literacy among workers can be insurmountable hurdles (Sony & Naik, 2019). Therefore, it is crucial to have ongoing training and development programs to keep employees up-to-date with the latest technology.

2.11.3 Economic impact on small- and medium-sized enterprises (SMEs) and large enterprises

Industry 4.0 is a phenomenon that manifests itself in different economic ways with regard to SMEs and large enterprises. Big enterprises can afford to use high-techs and training resources, while SMEs may suffer from financial constraints. Even if SMEs cannot catch up with the progress of large enterprises fully, the adoption of Industry 4.0 still brings great profits to the SME through such as flexible production, cost reduction, and improvement of competitiveness (Moeuf et al., 2018). Governments and policymakers are crucial for aiding SMEs to capital in resistance to arise that spend, incentives, and law that patronize the exercise of digital technology (Janssen et al., 2020).

2.11.4 Sustainability and circular economy

The integration of circular economy into the manufacturing processes is another way that Industry 4.0 promotes sustainability. IoT technology as well as blockchain supports tracking and tracing materials and makes recycling easier and wastes less (Antikainen et al., 2018). Industry 4.0 aims for resource efficiency and environmental compatibility and is referred to as being aligned with the principal goals of sustainability (Pagoropoulos et al., 2017).

2.11.5 Future research and trends

In the future, advancements in smart manufacturing will be led by the ongoing development of Industry 4.0 technologies. The research will help to develop better assessment frameworks for Industry 4.0 readiness, especially for SMEs. Moreover, an exploration of the fusion of Industry 4.0 with the next generation of technological advances (5G, augmented reality (AR), and blockchain) presents a range of new possible avenues for advancement (Xu et al., 2018). Academia, industry, and policymakers have to team up to tackle these challenges and to utilize the full potential of Industry 4.0.

2.12 CONCLUSION

The move to Industry 4.0 affects a major shift in the nature of manufacturing, enabling production improvements in efficiency, flexibility, and sustainability. But addressing the challenges that come with digital overuse is going to demand an all-hands-on-deck strategy that incorporates innovation, policy changes, new business models, workforce training, and updated laws. This is a reminder that as we continue the digital transformation research efforts, it is a collective pursuit we are a part of to maximize the economic potential of Industry 4.0.

REFERENCES

Antikainen, M., Uusitalo, T., & Kivikytö-Reponen, P. (2018). Digitalization as an enabler of circular economy. *Procedia CIRP, 73*, 45–49. https://doi.org/10.1016/j.procir.2018.04.027

Hermann, M., Pentek, T., & Otto, B. (2016). *Design principles for Industrie 4.0 scenarios: A literature review*. Technische Universität Dortmund.

Janssen, R., Liu, C., & Boorsma, M. (2020). *Cybersecurity in Industry 4.0: The Industrial Internet of Things*. CRC Press.

Kagermann, H., Wahlster, W., & Helbig, J. (2013). *Recommendations for implementing the strategic initiative INDUSTRIE 4.0. Securing the future of German manufacturing industry. Final report of the Industrie 4.0 Working Group*. Forschungsunion.

Lasi, H., Fettke, P., Kemper, H. G., Feld, T., & Hoffmann, M. (2014). Industry 4.0. *Business & Information Systems Engineering*, *6*(4), 239–242. https://doi.org/10.1007/s12599-014-0334-4

Mittal, S., Khan, M. A., Romero, D., & Wuest, T. (2018). A critical review of smart manufacturing & Industry 4.0 maturity models: Implications for small and medium-sized enterprises (SMEs). *Journal of Manufacturing Systems, 49*, 194–214. https://doi.org/10.1016/j.jmsy.2018.10.005

Moeuf, A., Pellerin, R., Lamouri, S., Tamayo-Giraldo, S., & Barbaray, R. (2018). The industrial management of SMEs in the era of Industry 4.0. *International Journal of Production Research*, *56*(3), 1118–1136. https://doi.org/10.1080/00207543.2017.1372647

Monostori, L., Kádár, B., Bauernhansl, T., Kondoh, S., Kumara, S. R. T., Reinhart, G., Sauer, O., Schuh, G., Sihn, W., & Ueda, K. (2016). Cyber-physical systems in manufacturing. *CIRP Annals, 65*(2), 621–641. https://doi.org/10.1016/j.cirp.2016.06.005

Negri, E., Fumagalli, L., & Macchi, M. (2017). A review of the roles of digital twin in CPS-based production systems. *Procedia Manufacturing, 11*, 939–948. https://doi.org/10.1016/j.promfg.2017.07.198

Pagoropoulos, A., Pigosso, D. C., & McAloone, T. C. (2017). The emergent role of digital technologies in the circular economy: A review. *Procedia CIRP, 64*, 19–24. https://doi.org/10.1016/j.procir.2017.03.178

Sanders, A., Elangeswaran, C., & Wulfsberg, J. P. (2016). Industry 4.0 implies lean manufacturing: Research activities in Industry 4.0 function as enablers for lean manufacturing. *Journal of Industrial Engineering and Management*, *9*(3), 811–833. https://doi.org/10.3926/jiem.1940

Schumacher, A., Erol, S., & Sihn, W. (2016). A maturity model for assessing Industry 4.0 readiness and maturity of manufacturing enterprises. *Procedia CIRP, 52*, 161–166. https://doi.org/10.1016/j.procir.2016.07.040

Sony, M., & Naik, S. (2019). Industry 4.0 integration with socio-technical systems theory: A systematic review and proposed theoretical model. *Technology in Society, 59*, 101197. https://doi.org/10.1016/j.techsoc.2019.04.002

Tao, F., Qi, Q., Liu, A., & Kusiak, A. (2018). Data-driven smart manufacturing. *Journal of Manufacturing Systems, 48*, 157–169. https://doi.org/10.1016/j.jmsy.2018.01.006

Villani, V., Pini, F., Leali, F., & Secchi, C. (2018). Survey on human-robot collaboration in industrial settings: Safety, intuitive interfaces and applications. *Mechatronics, 55*, 248–266. https://doi.org/10.1016/j.mechatronics.2018.02.009

Xu, L. D., Xu, E. L., & Li, L. (2018). Industry 4.0: State of the art and future trends. *International Journal of Production Research*, *56*(8), 2941–2962. https://doi.org/10.1080/00207543.2018.1444806

Zhou, K., Liu, T., & Zhou, L. (2015). Industry 4.0: Towards future industrial opportunities and challenges. In *2015 12th international conference on fuzzy systems and knowledge discovery (FSKD)* (pp. 2147–2152). IEEE. https://doi.org/10.1109/FSKD.2015.7382284

Chapter 3

Smart manufacturing

Paranitharan K.P., Raja Chandra Sekar M., Ramesh Babu T., Balaji V., Thanushkar S., and Sree Ramnath Sudalai S.

3.1 THE SMART MANUFACTURING CONCEPT

Smart manufacturing (SM) is an advanced approach that leverages cutting-edge technologies such as Internet of Things (IoT), automation, and cloud computing–data analytics to enhance processes involved in manufacturing performance. The goal is to make manufacturing more efficient and responsive to changing demands. A smart factory is a facility that embodies and implements these principles; big data, Industrial Internet of Things (IIoT) devices, and connected worker platforms are all integrated into manufacturing facilities [1]. In Figure 3.1, researchers and industries around the globe illustrate how intelligence is applied to manufacturing. Many researchers and practitioners refer to SM as intelligent manufacturing (IM). Major nations have been addressing the necessity of modernizing and changing their manufacturing sectors in recent years, drawing society's attention to networking, digitalization, and smartness/intelligence in their production. Manufacturers need to drive peak profitability within their core business [2,3]. This requires strategies to cooperate with company strategies. The term "Industry 4.0" alludes to the display slant in mechanization, observing, and information mining from generation forms, now and then known as the fourth mechanical insurgency. IoT, cyber-physical systems (CPSs), Cloud Platforms (CP), Cognitive Computing (CC), and augmented reality/virtual reality (AR/VR) devices are the leading technologies in these areas [4].

This poses a considerable challenge for the Industry 4.0 framework as it strives to achieve production efficiency with minimal costs while accommodating high levels of customization. Production procedures, goods, and manufacturing systems are all ready for full digitization in the current environment, utilizing the right technologies to begin the digitalization process [5]. The two main groups are broadly classified into the following: First, it is designed for permanent installation in the product; hence, cheapness should be given priority to the second machines that monitor operations within the manufacturing line, which underscores effectiveness in synchronization in performance measures.

DOI: 10.1201/9781003470861-3

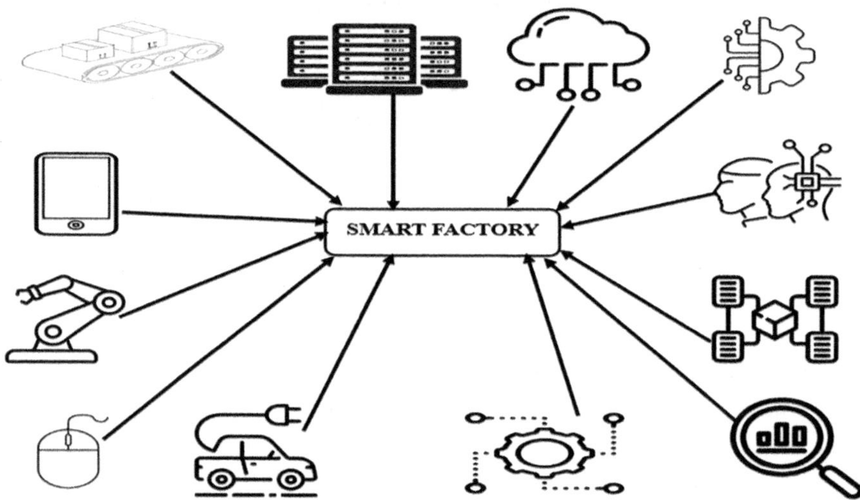

Figure 3.1 The smart manufacturing.

In decision-making, SM is linked to infrastructure-related and manufacturing outputs, such as cost, quality, delivery, flexibility, and innovation. Organizations provide a connection between manufacturing decision inputs, outputs, and competitive priorities that will lead to the best production of smart performance. This approach provides a way forward for practices on successfully deploying and managing smart manufacturing systems (SMSs), formulating an outcome-measuring framework. Digital manufacturing tools and techniques further optimize production workflows, while additive manufacturing opens doors to new possibilities in product design and creation. Communication ensures lightning-fast data exchange while robotics and automation take on increasingly complex tasks. Sophisticated big data processing and analytics extract insightful information from the sea of data created. Finally, system integration and flexible manufacturing systems (FMS) ensure seamless adaptability and responsiveness to dynamic market shifts [6].

In essence, SM is not just a technological upgrade; it's a complete paradigm shift. It's about embracing the power of data and connectivity to create a more agile, efficient, and customer-centric manufacturing landscape. It's about building factories that react, learn, and adapt, ultimately delivering the products and services consumers crave in the interconnected SMS with I4.0. The integration is facilitated through cloud computing, enhancing the efficiency of product personalization, on-demand production, and the general control of the demand and supply chain [7].

3.2 DRIVERS FOR ADOPTION OF SMART MANUFACTURING IN LARGE AND SMALL OR MEDIUM ENTERPRISES

Many reasons explain why SM has gained ground among large entities and small- and medium-sized enterprises (SMEs). The primary factor lies in the necessity of higher operational efficiency. The evolutionary field in modern industries is pushing enterprises to enhance efficacy in their production process to keep up with competitors. Consequently, flexible and adaptable production systems can quickly respond to ever-shifting consumer needs. SM also results from the desire for sound management and decision-making processes [8]. Enterprises will use the data collected in real time from different points of the production cycle, thus informing them on how to decide strategically. This is especially important within large enterprises that run complicated operations because real-time insights help coordinate actions between different business functions, leading to an organization's more uniform strategic direction [9]. Smart manufacturing is also driven by cost reduction and minimization of resources applied. These systems incorporate predictive maintenance features, allowing organizations to take a proactive approach towards addressing equipment problems, ensuring minimum time intervals and reduced maintenance expenses. Additionally, in recent times, companies have adopted smart manufacturing due to global change for sustainable and environmentally friendly practices. This will also help monitor and optimize waste, enhance resource efficiencies, and align with evolving corporate social responsibilities and green business practices [10,11].

The primary objective is to explore the decision considerations about including smart factories in SME manufacturing settings. In addition, it tries to empirically examine these decision elements within the domain of the manufacturing sector by using figures from Korean SMEs. This SM concentrates on identifying the influence of these choice components on the predisposition toward smart factory development and different phases of implementation. This understanding should be about all these determinants, including their small- and medium-sized business management implications and whether they have adopted or not yet decided to embrace the smart factory concept. Therefore, the overall goal of this research is to equip SMEs with the information they need to make rational decisions [12,13].

Table 3.1 shows that exploring SM technologies reveals a diverse landscape often described using varied terminology and levels of granularity of drivers across the literature. A consolidation effort aims to cluster related technologies and characteristics based on semantic similarity to address this. The above enabling factors and the guidelines for successful SM implementation are identified from existing literature. Notably, these factors may not always be required together. The resultant consolidation seeks to streamline

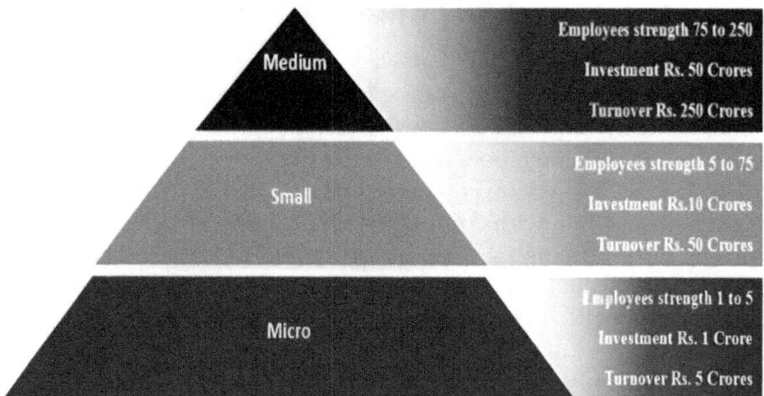

Figure 3.2 Small and medium enterprises (SMEs).

Table 3.1 Adaptation of drivers in smart manufacturing

S.No.	*Drivers*	*Source*
1	Sustainable energy	[15,16,17,18]
2	Environment-friendly	[19]
3	Law and regulations	[20]
4	Innovative education and training	[21]
5	Traditional environment	[22,23]
6	Digitalization	[24,25,26]
7	Enterprise integration	[27]
8	Core manufacturing simulation data	[28]
9	Strategic vision	[29]
10	Commitment to inventiveness	[30]
11	Cross-functional collaboration	[31]
12	Talent development and training	[32]
13	Risk management	[33]
14	Performance metrics and KPIs	[34]
15	Supply chain optimization	[35]
16	Knowledge workers	[36]

understanding amid the nuanced terminology and varying levels of detail prevalent in SM discourse. The SM aims to enlighten the public on whether or not modern strategies are concurrent with the vision that policymakers and scholars have established for small-scale manufacturing enterprises embracing smart factories. This is an extensive search that ultimately helps strategically incorporate smart factories into the manufacturing area for the different stakeholders [14].

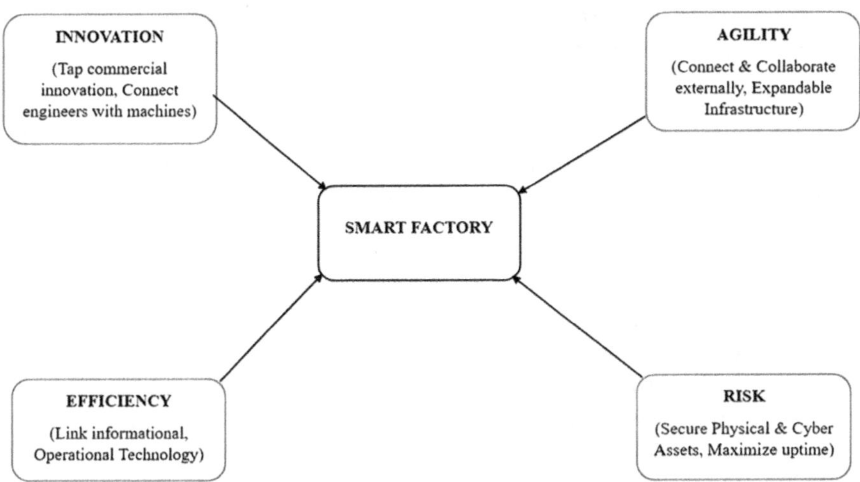

Figure 3.3 Smart factory.

Through digitalization and intelligence techniques, many of these advancements seek to make systems smarter and more intelligent. Numerous data are produced by SMS. However, the amount and usefulness of those data differ according to the organization's size, scope, and activities [20,37]. Computer-aided design (CAD) coupled with hybrid prototyping techniques like augmented reality and virtual reality are central to the emergence of intelligent design architecture in a production setting. Synergistically, this creates cyberspace in a CPS. An additional step in advancing the manufacturing paradigm is eliminating the necessity to conduct quality checks after the completion of a process. As stated, through the integration of the SMS transforms into an autonomous control system. Integration of the system improves its learning capability while reducing the frequency of manual for optimized efficiency of operations. The Smart factor will discuss how such real-time monitoring empowers the system to react dynamically to the operational sources, thus improving overall operational quality in SMEs, as shown in Figure 3.3. However, at the same time, it reduces the failure rate of products and equipment, which points to the system's integrity [38,39]. Also, the SMS is great at creating perfect scheduling systems, which are crucial for smooth operations. Simply in order to create a SMS, several contemporary technologies are combined, including cloud computing, machine learning, IIOT, CAD, augmented reality, and virtual reality. The system is revolutionary in design procedures, but it also underpins an encompassing transformation of manufacturing activities to promote efficiency, responsiveness, and dependability to become a smart factory [40,41].

3.3 BENEFITS OF THE ADOPTION OF SMART MANUFACTURING IN LARGE AND SMALL OR MEDIUM ENTERPRISES

While larger companies have already made progress in implementing SM, medium enterprises (SMEs) face challenges when adopting SM and creating a roadmap for its implementation. To aid small- and medium-sized manufacturing enterprises in adopting SM, we develop and evaluate a framework tailored to meet the needs of SMEs. Through a study involving SMEs that have already embraced SM, framework for SMEs considering smart manufacturing has been proposed. Adoption: first, identify the manufacturing data that are available; second, assess the readiness of SME data management processes; third, raise awareness about SM among leaders and employees within the SME; fourth, create a customized vision for how SM can benefit the needs of the SME; fifth, select appropriate tools and practices to bring this tailored vision to life [42].

SMSs now outsource conventional manufacturing, which various organizations essentially embrace for better performance. The development of SMS is based on a combination of using cutting-edge technologies such as artificial intelligence (AI), automation, data interchange, CPSs, IoT, and automated industrial systems, which gives it a high level of complexity and cost. SMEs with limited financial resources strive to ensure that any potential rewards are commensurate with adopting SMS. Using exploratory and empirical research, it identifies and proves measures of success of SMS investment in SMSs' Indian auto component manufacturing [43,44]. To achieve economic development in Asia, apart from large companies, SMEs are putting fourth industrial revolution strategies into practice to maximize performance. Industry 4.0 is the fourth generation of the industry. Combining various technologies and applications can convert traditional manufacturing systems into smart ones. The use of big data and its quantity among sectors, companies' sizes, and processes depend on SMS. Designing, machining, monitoring, scheduling, and controlling are included in SMS applications. SMS means without post-process inspection and with sensors, the self-enhancement of IoT, machine learning, and cloud computing. Scheduled optimization, reduced failure rates, and high-quality operation are all made possible by IoT in SMS through remote monitoring and control [45].

3.4 IMPACT OF SMART MANUFACTURING ON SUSTAINABLE DEVELOPMENT

SM has far-reaching effects on achieving sustainable development in many economic, social, and environmental aspects. An important aspect involves resource efficiency, which incorporates modern technologies such as IIoT and data analysis for accurate tracking and optimization of their use. In addition to minimizing waste, it also makes the whole process less bulky;

thus, it aligns with sustainable concepts. SM further enables the making of more sustainable products through design improvement and environmentally friendly materials [15,46]. SM offers better connectivity and enhances the agility of supply chain management through the reduction of the environmental footprint attributed to transport and logistics.

Smart manufacturing facilitates the emergence of new jobs that address social challenges relating to technological advancements, innovation and data analysis, economic transformation, and community well-being. It becomes an essential enabler by improving industrial practices toward sustainable development. It concerns the product life cycle, from the first conception through the whole production process and the phase of end-of-life [16]. Industry 4.0 is anticipated to promote cleaner energy and material resources and decrease waste in value-creation processes, which would help the environmental aspect of manufacturing sustainability.

In this regard, renewable energies in smart manufacturing are part of sustainable energy objectives, improving the environment for industrial actions. Also, smart manufacturing encourages the reduction of the traditional linear production models and hence adopts the circular economy approach [17]. Figure 3.4 shows an economical utilization of resources;

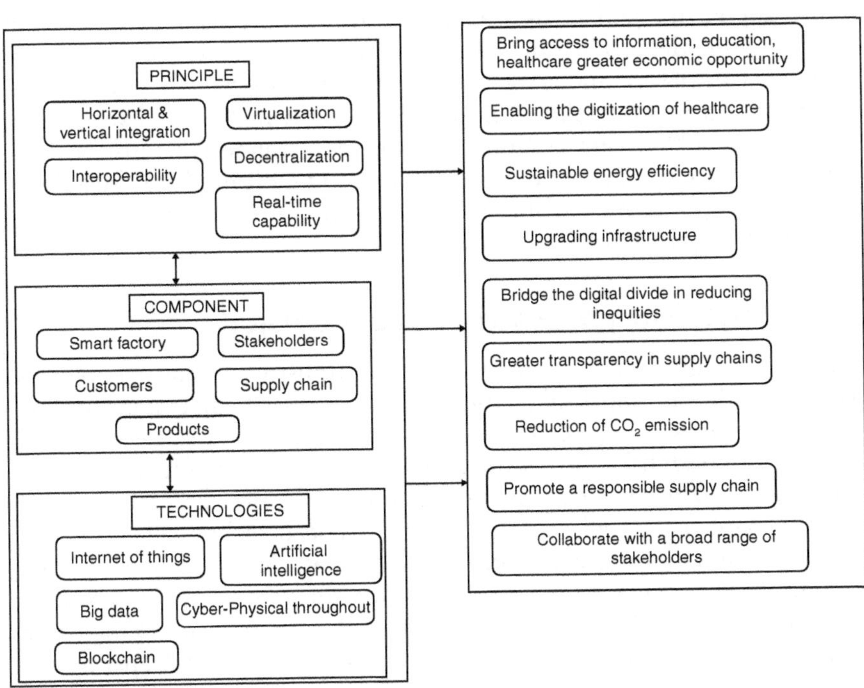

Figure 3.4 Sustainable development through smart manufacturing.

re-using and minimizing wastage in the process promotes a sustainable approach. It also provides specific tracking for environmental performance; such an open system promotes compliance with environmental rules and encourages a healthy corporate environment. Accessible real-time insights aid key players such as consumers and regulatory bodies in making proper decisions aimed at sustainability. The sustenance is built on the notion that it was developed from Industry 4.0, which marked a growth in the fourth wave of the Industrial Revolution [18,47]. Sustainable development is achieved by integrating the technology with respective functions in sustainability practice and sustainable manufacturing to achieve sustainability.

Sustainability 4.0 is a societal philosophy that frames issues such as society, ecology, and ethics beyond pure technological development. Digitalization has become an organizational principle, suggesting that the application of digital embedded technology is part of the essence of the organization. It involves combined value-creation procedures in firms and amongst partner organizations [22,23]. Further, there is an emphasis on decentralization structure and the use of cloud-based organizations to form. In a nutshell, issues regarding sustainable development must be addressed when defining Sustainability 4.0. As the IoT continues to improve, businesses must select solutions that will allow them to understand the amount of carbon dioxide generated and the resulting environmental effects in different procedures. Moving towards environmental responsibility will go a long way in supporting global sustainability objectives. In this regard, it centers on stimulating the production side, for instance, by creating jobs, providing safety at workplaces, and actively fighting against forced and child labor in factories [24]. Sustainability 4.0, as it refers to the inclusion of social issues in the main of the issues addressed by the contemporary industrial environment, is about achieving a well-balanced and ethically based sustainable development program, as opposed to the one that focuses only on economic considerations [25,26].

3.5 TRANSITION TO SMART MANUFACTURING

Our societies and businesses will continue to transform in the coming years, particularly in the industrial sector, resulting in even greater changes than in the past few decades. The change occurs due to the factory's deep embeddedness of ICT and fabrication technologies. This involves challenges related to competition in the business and social and environmental concerns [48]. This is enabled by CPS, which adopts advanced technologies, including additive manufacturing, collaborative robots, and virtual/augmented reality, to contribute to high-technology manufacturing procedures [49,50]. CPS is applied to manufacturing to promote self-organization, context-aware control, symbiotic human-robot collaboration, and the transformation of existing factories into futuristic ones. The elements that typify SM and smart factories define this transition [51].

The pandemic-accelerated development of such practices highlights the need for sustainable and flexible production. The adoption trends of global SM technology, such as cybersecurity, cloud connectivity, IoT, and machine vision, are 28%, while in North America, the rest are others. Globally, it is still at its infant stage, and the pandemic presents the need for speed in embracing SM through accessible implementation areas with minimal investment. It can be hindered by a deficiency of awareness, limited financial resources, and inadequate skills, necessitating an individualized adoption strategy [52]. There is some scholarly work on SME maturity and preparedness, but there is a lack of suitably selected, intelligent production technologies adapted for SME needs of smart sustainable systems [53].

SM could potentially change manufacturing in large and small companies. Nevertheless, manufacturers can use insufficient decision-making tools to measure the benefits and costs of implementing such technologies. Figure 3.5 shows the transition to a SM process [54].

Using IoT sensors and services linked to machinery, an SMS identifies issues with manufacturing opportunities for automated activities [55]. Technology integration in advanced stages and strategic initiatives is important in various aspects like the procurement of IoT devices, investment in automation, implementation of cybersecurity measures, and employee training. Such a multi-faceted approach enables manufacturers to fully exploit SM by enhancing efficiency, quality, flexibility, and competitiveness in today's volatile industrial landscape. This allows information to flow freely between interconnected systems, enabling real-time decisions and optimization of production processes through data collection from IoT sensors, sophisticated analytics, and AI algorithms. Radio Frequency Identification (RFID) tags that monitor the movement of incoming raw materials through the supply chain logistics, such as production processes, are just a few examples of automation, IoT, or data-driven decision-making leading to material flow optimization. Thus, such an inclusive strategy has

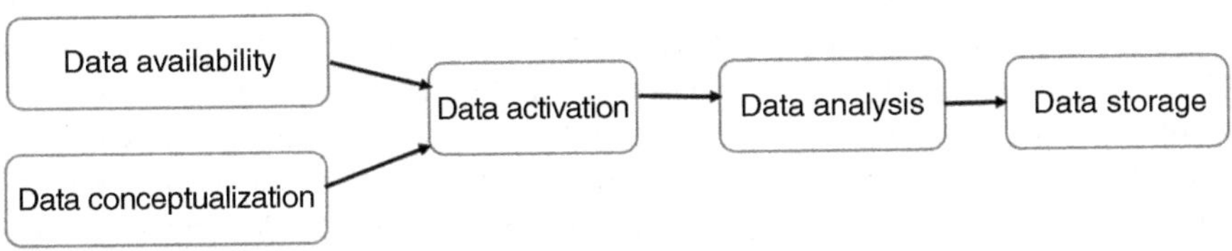

Figure 3.5 Transition to smart manufacturing.

led to efficiency and innovation and enabled firms to be future-oriented toward the digital economy.

The system process involves making the most of sophisticated technology using data-driven methods to increase efficiency, quality, and flexibility. The design framework emphasizes incorporating IoT, AI, and automation in product designs and manufacturing processes with sustainability in mind, so it is used to assess the preferences of many industry standards for implementing digital and information technologies in SMSs for Industry 4.0 that are sustainable [56]. Planning entails strategic and tactical optimization involving real-time data analysis, leading to agility and continuous improvement. Industry standards are upheld for interoperability, dependable communication between devices (reliability) and security, while a comprehensive knowledge base drives informed decision-making and innovation. In automated production systems, flexible manufacturing processes using adaptive algorithms based on AI, which can be made possible through IoT sensors, are examples of such capabilities in today's industry. Regular technology checks guarantee the effective operation of manufacturing systems, including cybersecurity, whereas data acquisition allows proactive maintenance and optimized planning. These include adjusting their operations automatically depending on real-time information obtained from several sources. In addition, simplified material handling methods have been effectively applied by micro-logistics optimization, resulting in improved productivity provision due to increased sensitivity (responsiveness) within factory areas. Efficiencies will, therefore, vary between different stages in the logistics implementation.

In SM, how data flows from one device to another is vital in ensuring that information is exchanged seamlessly and used intelligently between different systems to enable real-time decision-making and process optimization. The beginning of this procedure entails data collection from sensors, machinery, and production lines using the IoT, which are then transmitted to central platforms or cloud-based repositories that are processed and analyzed by employing advanced analytics and AI algorithms for this Big data [57] can revolutionize manufacturing into SM. In other words, these insights form the basis for making decisions that can be implemented, thereby encouraging partnership, openness, and quickness across all production areas. Concurrently, product flow optimization mechanizes the movement of materials and components throughout production using modern technologies that provide accuracy and adaptability. Moreover, cloud computing offers expandable, safe environments where data may be kept, processed, or scrutinized to provide continuous visibility into production processes and insight-driven collaboration leading to constant improvement or enhancement of those activities. Consequently, the transition to SM improves efficiency, quality, and competitiveness in today's ever-changing industrial scenario.

3.6 INTELLIGENT AUTOMATION SYSTEM

For strategic viewpoints on intelligent automation, academic research is a useful source of direction. A multitude of research studies, many utilizing reliable, rigorous methodologies, have examined the possible effects of AI on the workplace. Nevertheless, because these contributions are rooted in different academic fields and rely on divergent research paradigms, theories, methodologies, and viewpoints, there isn't agreement on important conclusions and their implications. It is in the best position for Information System researchers to put together a comprehensive knowledge of this new research issue because they operate at the nexus of numerous academic fields, taking into account both social and technical factors. Since the computer revolution of the 20th century, economists have used "computerization" to replace human labor with computers in job fulfillment.

Expanding on this concept, Frey and Osborne define computerization as the "automation of jobs through computer-controlled equipment, such as machine learning and mobile robotics," embodying the most recent progress in AI technology. Intelligent automation systems, encompassing AI and machine learning technologies, require substantial information to operate effectively [53]. These systems are designed to mimic human cognitive functions, enabling them to analyze data, make decisions, and perform tasks autonomously. They rely on comprehensive datasets as the foundation for learning and adaptation to function optimally. This information can include historical data, patterns, and various inputs relevant to the task or domain the automation system addresses. Intelligent automation systems demand frequent updates and closed-loop designs to fine-tune their algorithms to deliver optimal outputs in subsequent iterations. Furthermore, the type and complexity of the information presented are instrumental in determining whether the system will make correct forecasts. Smart automation systems need to integrate domain-specific knowledge. The knowledge can range from industry expertise to regulatory frameworks and contextual information to increase the systems' comprehension of the environment in which an application is running. As the system becomes more intricate and situational, it can handle sophisticated situations and make smart conclusions. The latest advancements in AI mark a distinct departure from DSS and knowledge-based systems, which were distinguished in three significant aspects during their initial stages. To begin with, these new systems do not require ongoing programming assistance from human programmers; instead, they may automatically learn from their experiences and improve both their procedures and results.

While the older systems worked alongside human professionals, offering suggestions and guidance, they still necessitated human involvement in the final decision-making. In the end, the traditional mechanisms were created to support managers in dealing with monotonous choices and intricate

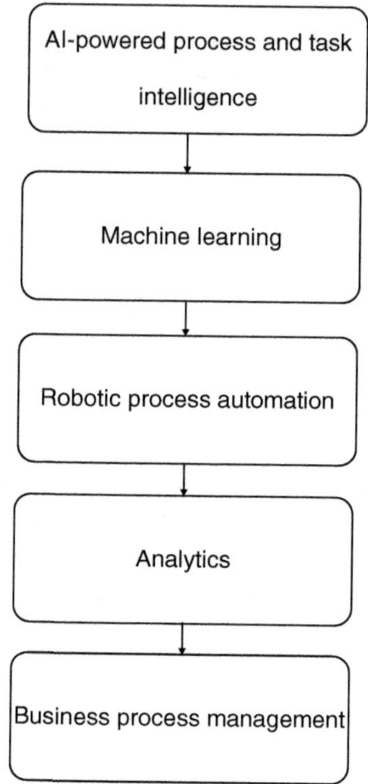

Figure 3.6 Intelligent automation system.

problems lacking a clear structure. Yet, they did not aim to alleviate mental activities from human workloads.

Figure 3.6 shows an AI representing an Intelligent Automation System, a significant development and broadening of the initial field of AI, bringing about a paradigm change in organizational dynamics compared to what was prevalent before [58]. The manufacturing industry relies heavily on adaptability and innovation. This advancement should result in sustainable production using new technology. To enhance sustainability, smart industrial technology must be viewed globally. As a result, the focus of today's IT giants has shifted to a number of AI subfields, including machine learning, natural language processing, graphics processing, and data mining. The subject of AI continues to capture significant attention within science due to the continuous advancement of current technologies [59]. Computer-integrated manufacturing (CIM) revolutionizes manufacturing by seamlessly integrating computers and automation across various functions, processes, and systems within an enterprise, from design to customer service. Key components such as CAD, Computer-Aided Manufacturing (CAM), Computer-Aided

Engineering (CAE), Product Lifecycle Management (PLM), and Enterprise Resource Planning (ERP) streamline operations [60], increase efficiency, and enhance flexibility. With CIM, you can monitor, analyze, and make decisions in real time, laying the foundation for SM. Hardware innovations like IoT devices, edge computing, and advanced robotics drive efficiency and quality, enabling real-time data collection, analysis, and optimization of manufacturing processes. Data analytics and simulation tools guide material selection and processing, ensuring optimal performance and reliability. Quality assurance is bolstered by sensors, IoT devices, and predictive maintenance techniques, minimizing defects and downtime. Decision and design tools such as digital twins and simulation software facilitate predictive maintenance, process optimization, and virtual prototyping, ultimately reducing costs and improving efficiency in SM environments.

Figure 3.7 lists a number of well-established research theories and techniques that have been documented over time, yet there hasn't been much integration or comparison between investigations. Considering the problem of limited resources and increasing output from the point of view of sustainability, the survey revealed a growing interest in applications for sustainable development and green manufacturing, demonstrating the critical role AI/ML plays in enhancing sustainability by making wise resource and energy decisions. The development of a new generation of IM, encompassing supply chain management, quality control, predictive maintenance, and energy consumption, are all aspects of sustainable processes that will be facilitated by the appropriate adoption of AI/ML technologies. The SMS responds dynamically to system status, customer needs, and supply chain networks [61].

An intelligent operator-machine system in SM integrates knowledge base, rule-based activities, skill-based activities, and AI to improve collaboration and performance. The knowledge base provides [62] operators with information and facilitates training. Rule-based activities ensure consistency and compliance – skill-based activities leverage operators' expertise. AI helps

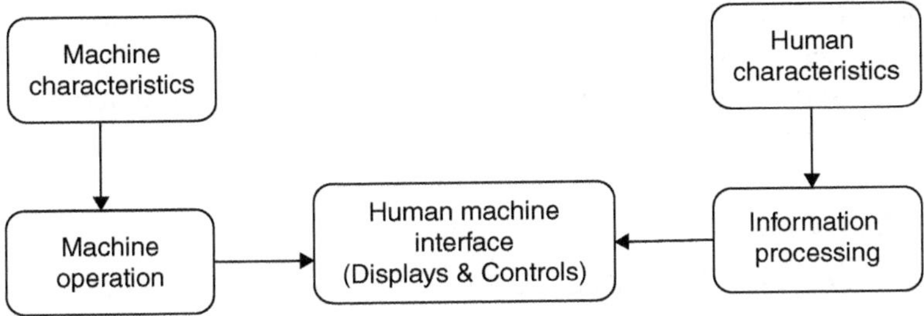

Figure 3.7 Intelligent operator-machine system.

with predictive maintenance, process optimization, decision-making, and automation. Optimal control algorithms optimize system performance based on real-time data. This system improves efficiency, productivity, and flexibility, achieving a competitive advantage in SM environments.

3.7 DEVELOPING SMART MANUFACTURING IN INDUSTRIES

Many AI systems use machine learning instead of explicitly programming a computer. Even though machine learning is often claimed to replicate human learning processes, it is simply the process of identifying patterns in data and then using those patterns to predict future events.

Models are fitted to training data despite the differences in fundamentals. Much like the robot mentioned earlier, machine learning relies heavily on the quality of its training data; a training set that encompasses all potential scenarios would be ideal [63].

While the crux of "digitally-driven production process innovation" may be captured from a macroperspective as innovation in production processes driven by digital technology, this approach paves the way for examination of the various adoption stages as well as factors leading to the technological innovation in smart factories from 1970 until now; examining the landscape of technology acceptance and innovation has been a primary research item among scholars. This search has pointed out some conditions that can be termed influential in adopting and implementing novel technologies [55,64]. Figure 3.3 explores how all these complicated ideas come together with technology adoption in a smart factory.

3.8 CONCLUSION

SM represents a transformative shift in the production industry by integrating advanced technologies such as the IoT, AI, big data analytics, and automation. The goal is to create more efficient, flexible, and sustainable manufacturing processes. By enhancing real-time decision-making, optimizing resource use, and improving product quality, SM is helping businesses achieve greater operational efficiency, reduce costs, and meet customer demands faster. As this approach continues to evolve, it holds the potential to revolutionize global supply chains, making manufacturing more resilient, adaptable, and competitive in the digital age.

REFERENCES

1. Abele, E., Kluge, J., & Näher, U. (2006). *Handbuch globale produktion* (p. 27). München: Hanser.

2. Wahlster, W. (Ed.). (2013). *SemProM: Foundations of semantic product memories for the Internet of Things*. Springer Berlin, Heidelberg: Springer Science & Business Media.
3. Wadhwani, D. (2023, May 5). *Industry 4.0 and the Benefits of Smart Manufacturing*. Smart Manufacturing. Retrieved December 22, 2024, from https://industry4o.com/2023/05/05/industry-4-0-the-benefits-of-smart-manufacturing/
4. Wright, P. (2014). Cyber-physical product manufacturing. *Manufacturing Letters*, *2*(2), 49–53.
5. Baheti, R., & Gill, H. (2011). Cyber-physical systems. *The Impact of Control Technology*, *12*(1), 161–166.
6. Vogel-Heuser, B., Bayrak, G., & Frank, U. (2011). Agenda CPS-Szenario smart factory. In *Erhöhte Verfügbarkeit und transparente Produktion. Tagungsband Automation Symposium* (pp. 6–21). Springer Vieweg, Berlin, Heidelberg.
7. Simmon, E., Kim, K. S., Subrahmanian, E., Lee, R., De Vaulx, F., Murakami, Y., ... & Sriram, R. D. (2013). *A vision of cyber-physical cloud computing for smart networked systems*. Gaithersburg, MD: US Department of Commerce, National Institute of Standards and Technology.
8. Lee, E. A., & Seshia, S. A. (2016). *Introduction to embedded systems: A cyber-physical systems approach*. Berkeley: MIT Press.
9. Rajkumar, R., Lee, I., Sha, L., & Stankovic, J. (2010, June). Cyber-physical systems: The next computing revolution. In *Proceedings of the 47th design automation conference* (pp. 731–736). ACM Digital Library.
10. Kagermann, H., Wahlster, W., & Helbig, J. (2013). *Recommendations for implementing the strategic initiative Industrie 4.0: Final report of the Industrie 4.0 Working Group (pp.* 12–25*)*. Berlin, Germany: Forschungsunion.
11. Wang, B., Tao, F., Fang, X., Liu, C., Liu, Y., & Freiheit, T. (2021). Smart manufacturing and intelligent manufacturing: A comparative review. *Engineering*, *7*(6), 738–757.
12. Židek, K., Piteľ, J., Adámek, M., Lazorík, P., & Hošovský, A. (2020). Digital twin of experimental smart manufacturing assembly system for Industry 4.0 concept. *Sustainability*, *12*(9), 3658.
13. Bernaert, M. (2012). Enterprise architecture for small and medium-sized enterprises. In *Confenis Doctoral Consortium Proceedings*. Presented at the Confenis Doctoral Consortium, Ghent, Belgium.
14. Phuyal, S., Bista, D., & Bista, R. (2020). Challenges, opportunities and future directions of smart manufacturing: A state of art review. *Sustainable Futures, 2*, 100023.
15. Al-Barakati, A., Mishra, A. R., Mardani, A., & Rani, P. (2022). An extended interval-valued Pythagorean fuzzy WASPAS method based on new similarity measures to evaluate the renewable energy sources. *Applied Soft Computing*, *120*, 108689.
16. Ali, H., Chen, T., & Hao, Y. (2021). Sustainable manufacturing practices, competitive capabilities, and sustainable performance: Moderating role of environmental regulations. *Sustainability*, *13*(18), 10051.
17. Ardolino, M., Rapaccini, M., Saccani, N., Gaiardelli, P., Crespi, G., & Ruggeri, C. (2018). The role of digital technologies for the service transformation of industrial companies. *International Journal of Production Research*, *56*(6), 2116–2132.

18. Ardanza, A., Moreno, A., Segura, Á., de la Cruz, M., & Aguinaga, D. (2019). Sustainable and flexible industrial human machine interfaces to support adaptable applications in the Industry 4.0 *paradigm. International Journal of Production Research, 57*(12), 4045–4059.
19. Ameta, K. L., Solanki, V. S., Singh, V., Devi, A. P., Chundawat, R. S., & Haque, S. (2022). Critical appraisal and systematic review of 3D & 4D printing in sustainable and environment-friendly smart manufacturing technologies. *Sustainable Materials and Technologies, 34*, e00481.
20. Mittal, S., Khan, M. A., Romero, D., & Wuest, T. (2018). A critical review of smart manufacturing & Industry 4.0 maturity models: Implications for small and medium-sized enterprises (SMEs). *Journal of Manufacturing Systems, 49*, 194–214.
21. Gajdzik, B., & Wolniak, R. (2022). Smart production workers in terms of creativity and innovation: The implication for open innovation. *Journal of Open Innovation: Technology, Market, and Complexity*, *8*(2), 68.
22. Bag, S., Gupta, S., & Kumar, S. (2021). Industry 4.0 adoption and 10R advance manufacturing capabilities for sustainable development. *International Journal of Production Economics*, *231*, 107844.
23. Bag, S., & Pretorius, J. H. C. (2022). Relationships between Industry 4.0, sustainable manufacturing and circular economy: Proposal of a research framework. *International Journal of Organizational Analysis*, *30*(4), 864–898.
24. Bag, S., Yadav, G., Wood, L. C., Dhamija, P., & Joshi, S. (2020). Industry 4.0 and the circular economy: Resource melioration in logistics. *Resources Policy*, *68*, 101776.
25. Banik, D., Ibne Hossain, N. U., Govindan, K., Nur, F., & Babski-Reeves, K. (2023). A decision support model for selecting unmanned aerial vehicle for medical supplies: Context of COVID-19 pandemic. *The International Journal of Logistics Management*, *34*(2), 473–496.
26. Bastas, A. (2021). Sustainable manufacturing technologies: A systematic review of latest trends and themes. *Sustainability*, *13*(8), 4271.
27. Uysal, M. P., & Mergen, A. E. (2021). Smart manufacturing in intelligent digital mesh: Integration of enterprise architecture and software product line engineering. *Journal of Industrial Information Integration, 22*, 100202.
28. Yao, X., Zhou, J., Lin, Y., Li, Y., Yu, H., & Liu, Y. (2019). Smart manufacturing based on cyber-physical systems and beyond. *Journal of Intelligent Manufacturing, 30*, 2805–2817.
29. Butt, J. (2020). A strategic roadmap for the manufacturing industry to implement Industry 4.0. *Designs*, *4*(2), 11.
30. Del Giudice, M., Scuotto, V., Papa, A., Tarba, S. Y., Bresciani, S., & Warkentin, M. (2021). A self-tuning model for smart manufacturing SMEs: Effects on digital innovation. *Journal of Product Innovation Management*, *38*(1), 68–89.
31. Arcidiacono, F., Ancarani, A., Di Mauro, C., & Schupp, F. (2022). The role of absorptive capacity in the adoption of smart manufacturing. *International Journal of Operations & Production Management*, *42*(6), 773–796.
32. Mittal, S., Khan, M. A., Purohit, J. K., Menon, K., Romero, D., & Wuest, T. (2020). A smart manufacturing adoption framework for SMEs. *International Journal of Production Research*, *58*(5), 1555–1573.

33. Jbair, M., Ahmad, B., Maple, C., & Harrison, R. (2022). Threat modelling for industrial cyber-physical systems in the era of smart manufacturing. *Computers in Industry, 137*, 103611.
34. Supekar, S. D., Graziano, D. J., Riddle, M. E., Nimbalkar, S. U., Das, S., Shehabi, A., & Cresko, J. (2019). A framework for quantifying energy and productivity benefits of smart manufacturing technologies. *Procedia CIRP, 80*, 699–704.
35. Oh, J., & Jeong, B. (2019). Tactical supply planning in smart manufacturing supply chain. *Robotics and Computer-Integrated Manufacturing, 55*, 217–233.
36. de Assis Dornelles, J., Ayala, N. F., & Frank, A. G. (2022). Smart working in Industry 4.0: How digital technologies enhance manufacturing workers' activities. *Computers & Industrial Engineering, 163*, 107804.
37. Won, J. Y., & Park, M. J. (2020). Smart factory adoption in small and medium-sized enterprises: Empirical evidence of manufacturing industry in Korea. *Technological Forecasting and Social Change, 157*, 120117.
38. Geissbauer, R., Vedso, J., & Schrauf, S. (2016). *Industry 4.0: Building the digital enterprise. PwC's 2016 Global Industry 4.0 Survey*. [Online]. www.pwc.com/gx/en/industries/industries-4.0/landing-page/industry-4.0-building-your-digital-enterprise-april-2016.pdf
39. Won, J. Y., & Park, M. J. (2020). Smart factory adoptio n in small and medium-sized enterprises: Empirical evidence of manufacturing industry in Korea. *Technological Forecasting and Social Change, 157*, 120117.
40. Lenz, J., MacDonald, E., Harik, R., & Wuest, T. (2020). Optimizing smart manufacturing systems by extending the smart products paradigm to the beginning of life. *Journal of Manufacturing Systems, 57*, 274–286.
41. Kavakli, E., Buenabad-Chávez, J., Tountopoulos, V., Loucopoulos, P., & Sakellariou, R. (2018, June). WiP: An architecture for disruption management in smart manufacturing. In *2018 IEEE international conference on smart computing (SMARTCOMP)* (pp. 279–281). Greece: IEEE.
42. Moghaddam, M., Cadavid, M. N., Kenley, C. R., & Deshmukh, A. V. (2018). Reference architectures for smart manufacturing: A critical review. *Journal of Manufacturing Systems, 49*, 215–225.
43. Abidi, M. H., Mohammed, M. K., & Alkhalefah, H. (2022). Predictive maintenance planning for Industry 4.0 using machine learning for sustainable manufacturing. *Sustainability, 14*(6), 3387.
44. Abubakr, M., Abbas, A. T., Tomaz, I., Soliman, M. S., Luqman, M., & Hegab, H. (2020). Sustainable and smart manufacturing: An integrated approach. *Sustainability, 12*(6), 2280.
45. Aggarwal, A., Gupta, S., Jamwal, A., Agrawal, R., Sharma, M., & Dangayach, G. S. (2022). Adoption of smart and sustainable manufacturing practices: An exploratory study of Indian manufacturing companies. *Proceedings of the Institution of Mechanical Engineers, Part B: Journal of Engineering Manufacture, 236*(5), 586–602.
46. Alayón, C. L., Säfsten, K., & Johansson, G. (2022). Barriers and enablers for the adoption of sustainable manufacturing by manufacturing SMEs. *Sustainability, 14*(4), 2364.

47. Ching, N. T., Ghobakhloo, M., Iranmanesh, M., Maroufkhani, P., & Asadi, S. (2022). Industry 4.0 applications for sustainable manufacturing: A systematic literature review and a roadmap to sustainable development. *Journal of Cleaner Production, 334*, 130133.
48. Bhandari, D., Singh, R. K., & Garg, S. K. (2019). Prioritising and evaluating barriers intensity for implementing cleaner technologies: Framework for sustainable production. *Resources, Conservation and Recycling, 146*, 156–167.
49. Bhanot, N., Rao, P. V., & Deshmukh, S. G. (2017). An integrated approach for analysing the enablers and barriers of sustainable manufacturing. *Journal of Cleaner Production, 142*, 4412–4439.
50. Bhatt, Y., Ghuman, K., & Dhir, A. (2020). Sustainable manufacturing. Bibliometrics and content analysis. *Journal of Cleaner Production, 260*, 120988.
51. Boral, S., Howard, I., Chaturvedi, S. K., McKee, K., & Naikan, V. N. A. (2020). A novel hybrid multi-criteria group decision making approach for failure mode and effect analysis: An essential requirement for sustainable manufacturing. *Sustainable Production and Consumption, 21*, 14–32.
52. Shao, G. (2021). *Use case scenarios for digital twin implementation based on ISO 23247*. Gaithersburg, MD: National Institute of Standards.
53. Lo, C. K., Chen, C. H., & Zhong, R. Y. (2021). A review of digital twin in product design and development. *Advanced Engineering Informatics, 48*, 101297.
54. Yuriy Skorenkyya, Roman Zolotyya, Sergiy Fedaka, Oleksandr Kramara, & Ruslan Kozaka (2023). *Digital Twin Implementation in Transition of Smart Manufacturing to Industry 5.0 Practices, CEUR Workshop Proceedings* (CEUR-WS.org).
55. Farahani, M. A., McCormick, M. R., Gianinny, R., Hudacheck, F., Harik, R., Liu, Z., & Wuest, T. (2023). Time-series pattern recognition in smart manufacturing systems: A literature review and ontology. *Journal of Manufacturing Systems, 69*, 208–241.
56. Didenko, N., Skripnuk, D., Kikkas, K., Kalinina, O., & Kosinski, E. (2021). The impact of digital transformation on the micrologistic system, and the open innovation in logistics. *Journal of Open Innovation: Technology, Market, and Complexity, 7*(2), 115.
57. Ochoa, W., Larrinaga, F., & Pérez, A. (2023). Context-aware workflow management for smart manufacturing: A literature review of semantic web-based approaches. *Future Generation Computer Systems, 145, 38–55.*
58. Tao, F., & Zhang, M. (2017). Digital twin shop-floor: A new shop-floor paradigm towards smart manufacturing. *IEEE Access, 5*, 20418–20427.
59. Surender, K., & Jha, A. K. *Computer-aided design and manufacturing*. New Delhi: Dhanpat Rai and Co. Ltd.
60. Yang, L., Zou, H., Shang, C., Ye, X., & Rani, P. (2023). Adoption of information and digital technologies for sustainable smart manufacturing systems for Industry 4.0 in small, medium, and micro enterprises (SMMEs). *Technological Forecasting and Social Change, 188*, 122308.
61. van Aart, C. J., Wielinga, B., & Schreiber, G. (2004). Organizational building blocks for design of distributed intelligent system. *International Journal of Human-Computer Studies, 61*(5), 567–599.

62. Tao, F., Qi, Q., Liu, A., & Kusiak, A. (2018). Data-driven smart manufacturing. *Journal of Manufacturing Systems, 48*, 157–169.
63. Kung, K. M., & Soepriyanto, G. (2020). To relocate or not to relocate? A case study of a Korean garment manufacturing company in Indonesia. *PalArch's Journal of Archaeology of Egypt/Egyptology, 17*(7), 2743–2754.
64. Ezell, S., & Dascoli, L. (2021). *How an information technology agreement 3.0 would bolster global economic growth and opportunity*. Washington, DC: Information Technology and Innovation Foundation.

Chapter 4

Advanced manufacturing systems and Industry 4.0

Natarajan Manikandan and Pasupuleti Thejasree

4.1 INTRODUCTION

All types of production processes rely on automation and supervision systems. These networks oversee the transportation, healthcare, water, energy, economics, and national security systems – all of which are vital to a well-functioning society [1]. Actuators change the process's behavior, while sensors record it. Communication with automation and control units, such as Programmable Logic Controllers (PLCs) in industry, is facilitated by them. On top of that, control and data acquisition systems provide for the real-time visualization of critical variables to track the process's development, making them an ideal supervisory and monitoring tool. These systems can create alarms in addition to providing numerical and graphical data. Furthermore, data is sent between the stated equipment through digital communication networks. Improvements in electronics, processing, communications, and control algorithms have been a part of the hardware and software development process since its inception in the 1970s [2–4]. Extensive use of automation technology in manufacturing processes was the principal force behind the third industrial revolution. Productivity, efficiency, traceability, dependability, and security are all improved by the use of real-time data monitoring, interchange, and gathering made possible by communication technology. Doing so will help keep expenses down and provide credence to the smart manufacturing idea [5]. As more and more businesses, processes, and infrastructures work to adopt the ideas and technologies of Industry 4.0 and the Industrial Internet of Things (IIoT) paradigms, these systems are becoming more and more important. Control and data acquisition systems are advancing the concepts of Industry 4.0 and the Industrial Internet of Things [6]. Classified as the "fourth industrial revolution," Industry 4.0 encompasses a plethora of cutting-edge technological developments. Internet of Things (IoT), cloud computing (CC), AI, and industrial cyber-physical systems (ICPSs) are all part of this category. The "Industry 4.0" concept was developed in 2009 as a component of the German government's Digital Agenda initiative. This idea, which is known

 DOI: 10.1201/9781003470861-4

as Industrie 4.0 in Germany, sparked a new age of technological innovation when it was unveiled at the 2011 Hanover Fair. A number of fields stand to benefit greatly from this paradigm shift, including those dealing with energy efficiency, sustainability, working environments, manpower, managing the production, maintenance planning, and many more. Also, this new condition has a direct impact on how automated and supervisory systems are implemented and operated. In addition to non-industrial processes, this merging concept's incorporation affects smart grids, smart cities, and similar initiatives [7]. Energy 4.0, Operator 4.0, Engineer 4.0, Education 4.0, and many more are examples of phrases that are either linked to the Industry 4.0 domain or marked with the number 4.0 to emphasize their innovative or advanced nature.

Undoubtedly, the literature demonstrates a growing number of publications focused on emerging technological advancements in sensing, data gathering, visualization, data storage, and analytics. PLC and Supervisory Control and Data Acquisition (SCADA) systems are also actively involved in these developments. Indeed, their existence and function in facilities that adhere to Industry 4.0 standards remain crucial and necessary [8–10]. In addition, an increasing number of technologies beyond basic automation and monitoring are being integrated into factories. More and more, industrial systems are integrating web-based interfaces, cybersecurity measures, IoT-enabled equipment, cloud data storage and processing, remote monitoring, and digital twins. In order to facilitate the integration of the aforementioned technologies with Industry 4.0, PLC and control and data acquisition systems have been enhanced and given new functions. Consequently, engineers versed in Industry 4.0 technologies are in high demand in the industry's job market. Informatics experts, software engineers, data analysts, cybersecurity experts, PLC programmers, and robot programmers are just a few of the many job profiles needed by the Industry 4.0 framework. The role of PLC programmers is particularly significant in this scenario [11,12]. A total of 100 new professional profiles are identified specifically for the purpose of developing future factories that are tailored to the requirements of Industry 4.0. The writers consist of a PLC programmer referred to as an Industry 4.0 PLC programmer, as well as industrial User Interface (UI) designers, specifically an industrial UI designer [13]. Similarly, in the realm of education, there is a steady increase in the number of training courses focused on Industry 4.0. This trend demonstrates a growing interest in these subjects. Indeed, higher education must address the problems and seize the opportunities presented by Industry 4.0 [14]. Academic institutions have long struggled with the task of producing industrial engineers, especially in light of Industry 4.0 [15]. Education and preparation of engineering students are of the utmost importance if we are to meet the challenges of the fourth industrial revolution, also known as Industry 4.0. They will be better able to solve problems after this. Training in critical technologies, including automation equipment, connectivity, and supervisory interfaces, is essential for engineers to operate efficiently in Industry 4.0 [16]. Modern technologies built in line with the principles of Industry 4.0 and the Industrial Internet of Things coexist with more traditional legacy equipment in today's economy. Consequently, it is essential for engineers and practitioners to be prepared to handle both kinds of challenges [17–20]. This study provides a comprehensive analysis of Industry 4.0, focusing on its idea, functional design, and recent changes, specifically in relation to automation and supervision systems. The secret passageway of the latest technological developments in software and hardware, following the trail from the vague idea of Industry 4.0, includes the shift from centralized automation designs centered on Industry 4.0 and IIoT to decentralized ones. The primary objective is to propose an inclusive overview of the principles and developments related to the convergence of Industry 4.0 and IIoT paradigms. Furthermore, we will elaborate on the impact of these paradigms on the equipment used for industrial automation and supervision, encompassing both hardware and software components. The present investigation develops an extensive reference document that will be valuable for professionals, engineers, and academics engaged in automation and supervision. This article presents a contextualization of Industry 4.0 as the 4th industrial revolution. It also offers many definitions and accompanying technologies related to this concept. Industry 4.0 and the IIoT are centered on a distributed and operational architecture, which is a shift from a hierarchical one in automation. The most recent evolution and patterns in the development of monitoring and automation systems that can be combined into infrastructure provided by Industry 4.0 have been presented through this research.

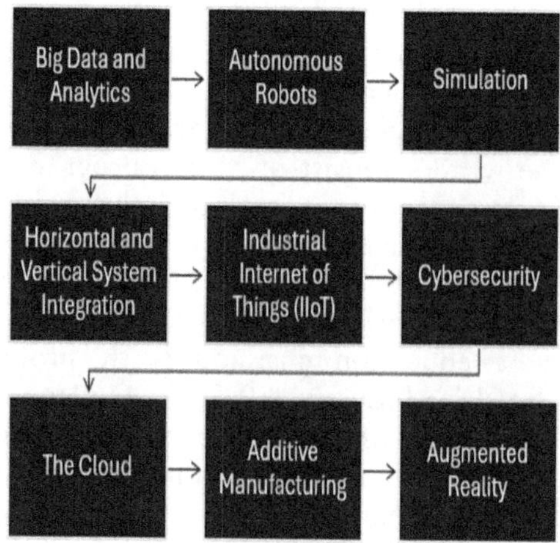

Figure 4.1 Pillars of Industry 4.0.

4.2 CONCEPT OF INDUSTRY 4.0

In order to be competitive, businesses must constantly adapt to new technology developments [21–23]. Managers should put a lot of effort into enhancing their company and production processes because of the rising level of competition in terms of productivity and quality. Many technologies currently exist that assist many businesses in improving performance and production; they include the IoT, Big Data, CC, digital twins, and additive manufacturing. "Industry 4.0" or "The Fourth Industrial Revolution" [24,25] refers to a broader notion that includes these technologies. The "Industrial Internet Consortium" in the United States and the "Industrial Value Chain Initiative" in Japan have all taken steps in this direction since 2011 when Germany unveiled a new strategic vector for industry development in the nation and introduced the "Plattform Industrie 4.0" [26]. The fourth industrial revolution, also known as Industry 4.0 (I-4.0), has the ability to alter production flow, human-machine communication, and the relationships among suppliers, manufacturers, and consumers. In addition to the aforementioned technologies, it incorporates autonomous robots, simulation, cybersecurity, augmented reality, horizontal and vertical system integration, and nine potential pillars that may be strengthened with AI solutions [27,28]. The foundation of Industry 4.0 is the idea of CPSs, which allow for the integration of virtual and physical systems. There is hope that the sector may reap operational, economic, and environmental benefits through the integration of big data, CC, artificial intelligence, and the IoT into automation and business operations. Machines and equipment in such a setup not only avoid centralized control systems by connecting to a single cloud, but they also acquire complete autonomy, allowing them to respond quickly to unforeseen situations. Various definitions of Industry 4.0 and associated ideas and technologies are covered in this section. With no universally accepted definition currently in place, we aim to provide a comprehensive view. Industry 4.0 is commonly thought of as the fourth industrial revolution; therefore, before we go into these criteria, we provide a quick historical summary of the preceding revolutions. Additionally, a list of related public and private initiatives is given, which emphasizes the clear interest that has grown.

4.2.1 History of revolutions in industry

Many people refer to Industry 4.0 as the Fourth Industrial Revolution. Putting it in its proper historical perspective will help us better grasp its significance. This is the accepted view on the subject of industrial revolutions, which might be anywhere from four to five in number. The First Industrial Revolution was greatly accelerated by the late 18th-century invention of the steam engine by James Watt [29]. This finding allowed for the utilization of steam-powered mechanical equipment in several sectors. Significant social and economic upheavals were also brought about by the technical implications. Everyone is talking about "Industry 1.0" at the moment. A whole new surge of production occurred between the late 1800s and the mid-1900s; this period is commonly known as the Second Industrial Revolution or Industry 2.0 [30]. During this time, the assembly line and other forms of mass manufacturing were greatly facilitated by the widespread use of electricity. The third industrial revolution, widely known as Industry 3.0 or the "Digital Revolution," began in the mid-century. Automation in manufacturing is defined by the usage of PLCs, or Programmable Logic Controllers, which were invented in 1969. In addition, industrial plants included the latest innovations in electronics, robotics, information technology, and telecommunications. PLCs were used to automate certain operations and collect or exchange data, which started the trend of digitizing factories [31,32]. The fourth industrial revolution, often known as Industry 4.0, has been underway since the beginning of the century. Uniting the digital, physical, and virtual worlds, it makes use of state-of-the-art technologies including 3D printing, complex materials, blockchain, robotics, IoT, nanotechnology, bioinformatics, and AI. A multitude of current technologies have brought about this transition, which might be characterized as revolutionary [33]. Emerging technologies and innovations are reaching more people and spreading at a faster rate than previous revolutions [34]. Full factory automation is one of the predicted results of Industry 4.0, which will be enabled by the extensive use of new technologies. As a result, cutting-edge automated production systems are now within reach [35]. By bringing attention to the need to update and modernize systems for remote work, the recent COVID-19 pandemic and lockdowns proved the significance of this revolution [36]. In order to achieve the aim of improved agility, resilience, and flexibility [37], which is closely related to Industry 4.0 and related technologies, it also highlighted the need of digital transformation.

4.2.2 Application of Industrial IoT in manufacturing

In order to facilitate the exchange and gathering of information and data, a network of physical items and devices that have sensors, software, and electronics is known as the IoT [38]. A manufacturing sector technology that allows products, equipment, and people to communicate with one another is the IIoT. Companies are installing sensors in machineries and other physical resources on the factory floor to gather data that can improve productivity and efficiency through real-time decision-making. The integration of several gadgets enables enhanced user experience and enables efficient decision-making. The potential benefits provided by the IIoT are illustrated in Figure 4.1. In order to take advantage of any of these chances, the most essential factor is the data. Therefore, it is necessary to adopt an efficient strategy for acquiring and consolidating data. Organizations necessitate dependable data to make well-informed judgments at each stage of the process. The utilization of IIoT frameworks offers a multitude of benefits to both consumers and industries. An essential characteristic of any sophisticated computer framework is the inherent reliability of codes and commands, which serve to prevent generally acknowledged problems and human mistakes. As a result, the reliability of many fundamental systems can be significantly enhanced. By integrating many autonomous systems, IIoT frameworks have the capability to transmit and interpret data at a level that is beyond human comprehension. By acquiring fragments of knowledge from a vast reservoir of information, it is now possible to achieve continuous advancements in productivity, scale, and performance. The IIoT facilitates the emergence of a new production approach known as personalized production, which allows for consumer engagement starting from the product design stage [39,40].

4.2.3 The importance of automation in manufacturing

The fourth industrial revolution, or Industry 4.0, has revolutionized the manufacturing sector by utilizing analytics, machine learning algorithms, and automation to streamline and automate repetitive activities. Human operators are now in charge of system maintenance and monitoring, leading to speedier and more efficient manufacturing processes. Wealthier nations may now compete with low-wage ones because of the enhanced competitiveness brought about by the adoption of contemporary technology. By producing more efficient and high-quality outputs and applying predictive and preventative maintenance and upgrades, Industry 4.0 boosts profits and income [33–35]. Growing labor costs in developed countries have prompted investments in automation as a means to rationalize the replacement of human workers with robots in manufacturing. Furthermore, automated procedures have been developed to replace human workers due to the shortage of labor in industrialized nations. Automating mundane and labor-intensive tasks has far-reaching social benefits and improves working conditions generally [36,37]. There has been a rise in worker safety regulations due to the fact that automation is taking over supervisorial roles. Occupational Safety and Health Act (OSHA) of 1970 places a strong focus on health and safety, and this is in line with that. By reducing human error and increasing consistency and adherence to quality standards, automation improves product quality. In addition, it shortens production lead times, giving businesses an edge by lowering the quantity of inventory in progress and the time it takes to get from an order to a completed product. Customer retention and regulatory compliance are both positively affected by better data tracking and analysis, which in turn improves record keeping. Industry 4.0's inherent closed-feedback loop, in contrast to more conventional methods of feedback, speeds up the feedback process for products and services. By encouraging better cooperation across the supply chain, greater analytics, collaborative data sharing, and improved connectivity improve production processes, leading to more productivity, more efficiency, and more innovation. Manufacturers, suppliers, and other value chain stakeholders may work together more effectively thanks to machine-to-machine linkages and integrated systems made possible by this integration. In conclusion, automation and data-driven processes are integral parts of Industry 4.0, which substantially improves manufacturing quality, competitiveness, and productivity [38–40].

4.2.4 Advancements in manufacturing automation in line with the principles of Industry 4.0

Known as the "fourth industrial revolution," the advent of Industry 4.0 has significantly altered the degree to which production processes are mechanized. Worldwide, manufacturing operations are becoming more efficient, productive, and adaptable as a result of the present revolution, which is defined by the integration of intelligent methodology and the IoT into several parts of industrial processes [1–8]. In order for firms to stay competitive in today's ever-changing business landscape, automation in manufacturing is a must. Commonly abbreviated as "Manufacturing 4.0" or "Industry 4.0," this trend describes the integration and improvement of digital technologies in production. Ex-Xplore Technologies CEO Mark Holleran calls it a complete shift from centralized to decentralized production. Changes to processes, employees, organizational structures, and technology are required to make this shift. Innovations in technology like digital manufacturing, smart sensors, CC, the IoT, artificial intelligence, and, most importantly, robots are ushering in the Fourth Industrial Revolution [9–12].

With the advent of Industry 4.0, manufacturing has undergone a sea shift, with automation playing an increasingly important role in enhancing overall performance and standardizing production processes. Traditional production methods are being revolutionized by automation technologies such as AI, machine learning, big data analytics, and robots. These advancements in technology make it easier for machines and systems to collaborate and share data, which in turn allows them to make better judgments in real time. The end outcome is improved accuracy, productivity, and economy of scale. Manufacturing automation optimizes production processes, which significantly increases efficiency. Manufacturing companies may improve their production efficiency by boosting throughput, decreasing cycle times, and eliminating mistakes by automating repetitive and time-consuming activities. Companies can easily meet the many needs and preferences of their customers thanks to automation, which allows for broad customization in production. In addition, automation can help make workplaces safer by lowering employees' exposure to potentially harmful tasks and environments [13–20]. There will be fewer accidents and injuries on the job thanks to robots and automated systems since they can undertake dangerous processes consistently and accurately. Because people can focus on higher-value tasks that require creativity, problem-solving, and decision-making talents, automation also leads to a more competent and efficient workforce. Industry 4.0 defines automation as permeating the whole industrial environment, not only the physical production area. Manufacturers may get rapid insight and command over their operations by merging CPSs, data analytics, and CC. The ability to optimize the supply chain, conduct demand-based production, and use predictive maintenance all contribute to a more nimble reaction to changes in the market. The benefits of factory automation are undeniable, but in order to reap the benefits of Industry 4.0, enterprises must overcome certain challenges. The initial costs, concerns about data security and privacy, and the need to train staff to run and maintain automated systems all constitute obstacles [21–26]. The long-term advantages of automation, including increased productivity, quality, and competitiveness, more than make up for the short-term drawbacks. As a whole, Industry 4.0's emphasis on production automation heralds a dramatic improvement in manufacturing's efficiency, flexibility, and impact on the environment. Adopting automation technologies and capitalizing on digitalization can open up new opportunities for development and innovation for companies. The complexities of the fourth industrial revolution will cause the role of automation in manufacturing to evolve, which in turn will cause many industries and nations to see economic development and advancement [27–32].

4.3 CONCLUSIONS

- Industry 4.0, also known as the 4th industrial revolution, has generated research efforts from both academia and industry, in addition to government programs.
- In order to stay competitive in the industry, it is essential to use cutting-edge technology and improve the capabilities of the workforce. Innovations in production techniques have resulted from the merging of digital and physical worlds, which were unthinkable even a decade ago. Thanks to Industry 4.0 techniques, this has become easier. However, there are still problems that must be fixed.
- Additional considerations are required after CPS and digital twin development. Quicker manufacturing and better personalization options are made possible by Industry 4.0's digitalized production.
- With advanced manufacturing systems (AMSs), manufacturers can print on-site, decrease waste, and enable customization, all while reducing transportation costs and time to market. As a result of the

COVID-19 pandemic and the enormous efforts of the AMS sector, the digitization of AMS will experience significant growth in the next years.

- Companies will have new options and possibilities opened up to them by this breakthrough. As the limitations of traditional production methods become more apparent, this will enable companies to react to real-time variations with greater speed and efficiency.
- In order to make the most of its potential and speed up the spread of Industry 4.0 in the manufacturing sphere, AMS must step up its research efforts; otherwise, it will continue to be at the forefront of technological advancements for the next several years. The paper evaluates relevant literature in order to determine the implications of implementing Industry 4.0.
- It is clear from the foregoing that Industry 4.0 may help industrial companies become more efficient and competitive. But the enormous expenses of installation, maintenance, and training are the key obstacles to the widespread use of Industry 4.0. Businesses will have an easier time implementing Industry 4.0 if they can convince their staff of the advantages of digital technology and give them confidence that it would be easy for them to use.

REFERENCES

1. Bahrin, M. A. K., Othman, M. F., Azli, N. H. N., & Talib, M. F. (2016). Industry 4.0: A review on industrial automation and robotic. *Jurnal teknologi*, *78*, 6–13.
2. Kagermann, H., Wahlster, W., & Helbig, J. (2013). Recommendations for implementing the strategic initiative Industrie 4.0: Final report of the Industrie 4.0 Working Group. *Acatech München*, *8*, 19–26.
3. Brettel, M., Friederichsen, N., Keller, M., & Rosenberg, M. (2014). How virtualization, decentralization and network building change the manufacturing landscape: An Industry 4.0 perspective. *FormaMente*, *8*(1), 37–44.
4. Morelli, G., Magazzino, C., Gurrieri, A. R., Pozzi, C., & Mele, M. (2022). Designing smart energy systems in an Industry 4.0 paradigm towards sustainable environment. *Sustainability*, *14*, 3315.
5. Sivakumar, K., Dhyankumar, C. T., Cherian, T. M., Manikandan, N., & Thejasree, P. (2023). Requirements for the adoption of Industry 4.0 in the sustainable manufacturing supply chain. In *Industry 4.0 technologies: Sustainable manufacturing supply chains: Volume II – Methods for transition and trends* (pp. 185–201). Singapore: Springer Nature Singapore.
6. Langmann, R., & Stiller, M. (2019). The PLC as a smart service in Industry 4.0 production systems. *Applied Sciences*, *9*, 3815.
7. Folgado, F. J., González, I., & Calderón, A. J. (2023). Data acquisition and monitoring system framed in Industrial Internet of Things for PEM hydrogen generators. *Internet Things*, *22*, 100795.

8. Tidrea, A., Korodi, A., & Silea, I. (2023). Elliptic curve cryptography considerations for securing automation and SCADA systems. *Sensors*, *23*, 2686.
9. Manikandan, N., Thejasree, P., Vimal, K. E. K., Sivakumar, K., & Kiruthika, J. (2023). Applications of artificial intelligence tools in advanced manufacturing. In *Industry 4.0 technologies: Sustainable manufacturing supply chains: Volume II – Methods for transition and trends* (pp. 29–42). Singapore: Springer Nature Singapore.
10. Pereira, A. G., Lima, T. M., & Santos, F. C. (2020). Industry 4.0 and Society 5.0: Opportunities and threats. *International Journal of Recent Technology and Engineering, 8*(5), 3305–3308.
11. Thejasree, P., Manikandan, N., Vimal, K. E. K., Sivakumar, K., & Krishnamachary, P. C. (2023). Applications of machine learning in supply chain management—A review. *Industry 4.0 technologies: Sustainable manufacturing supply chains: Volume 1—Theory, challenges, and opportunity* (pp. 73–82). Springer Nature.
12. Kumar, K., Zindani, D., & Davim, J. P. (2019). *Industry 4.0: Developments towards the fourth industrial revolution*. Cham: Springer.
13. Gadre, M., & Deoskar, A. (2020). Industry 4.0 – Digital transformation, challenges and benefits. *International Journal of Future Generation Communication and Networking*, *13*(2), 139–149.
14. Wan, J., Cai, H., & Zhou, K. (2015, January). Industrie 4.0: Enabling technologies. In *Proceedings of 2015 international conference on intelligent computing and internet of things* (pp. 135–140). Piscataway, NJ: IEEE.
15. Aghenta, L. O., & Iqbal, M.T. (2019). Low-cost, open source IoT-based SCADA system design using Thinger.IO and ESP32 thing. *Electronics*, *8*, 822.
16. Sverko, M., Grbac, T. G., & Mikuc, M. (2022). SCADA systems with focus on continuous manufacturing and steel industry: A survey on architectures, standards, challenges and Industry 5.0. *IEEE Access*, *10*, 109395–109430.
17. Babayigit, B., & Abubaker, M. (2023). Industrial Internet of Things: A review of improvements over traditional SCADA systems for industrial automation. *IEEE Systems Journal*, *18*(1), 120–133.
18. Hsiao, C. H., & Lee, W. P. (2021). *OPIIoT: Design and implementation of an open communication protocol platform for Industrial Internet of Things. Internet Things*, *16*, 100441.
19. Benešová, A., & Tupa, J. (2017). Requirements for education and qualification of people in Industry 4.0. *Procedia Manufacturing*, *11*, 2195–2202.
20. Sakurada, L., Geraldes, C. A. S., Fernandes, F. P., Pontes, J., & Leitão, P. (2021). Analysis of new job profiles for the factory of the future. In *Service oriented, holonic and multi-agent manufacturing systems for industry of the future: Proceedings of SOHOMA 2020* (Vol. 952, pp. 262–273). Berlin/Heidelberg: Springer Science and Business Media Deutschland GmbH: Berlin/Heidelberg.
21. Jiménez López, E., Cuenca Jiménez, F., Luna Sandoval, G., Ochoa Estrella, F. J., Maciel Monteón, M. A., Muñoz, F., & Limón Leyva, P. A. Technical considerations for the conformation of specific competences in mechatronic engineers in the context of Industry 4.0 and 5.0. *Processes*, *10*, 1445.
22. Benis, A., Nelke, S. A., & Winokur, M. *Training the next industrial engineers and managers about Industry 4.0: A case study about challenges and opportunities in the COVID-19 era. Sensors*, *21*, 2905.

23. Simons, S., Abé, P., & Neser, S. (2017). Learning in the AutFab—The fully automated Industrie 4.0 Learning Factory of the University of Applied Sciences Darmstadt. *Procedia Manufacturing*, *9*, 81–88.
24. IBM. (2024, October 9). Industria 4.0. *What is Industry 4.0?* Retrieved December 22, 2024, from www.ibm.com/es-es/topics/industry-4-0
25. Lee, M. H., Yun, J. H. J., Pyka, A., Won, D. K., Kodama, F., Schiuma, G., Park, H. S., Jeon, J., Park, K. B., Jung, K. H., et al. (2018). How to respond to the Fourth Industrial Revolution, or the Second Information Technology Revolution? Dynamic new combinations between technology, market, and society through open innovation. *Journal of Open Innovation: Technology, Market, and Complexity*, *4*, 21.
26. Folgado, G., Francisco, J., David, C.G., Isaías, G.P., &Antonio, J.C.G. (2024). "Review of Industry 4.0 from the Perspective of Automation and Supervision Systems: Definitions, Architectures and Recent Trends."
27. Schwab, K. (2016). *The Fourth Industrial Revolution*. Geneva: World Economic of Forum. ISBN 9781944835019.
28. ISO Smart Manufacturing Coordinating Committee White Paper on Smart Manufacturing. www.iso. org/files/live/sites/isoorg/files/store/en/PUB100459. pdf
29. Åkerberg, J., Åkesson, J. F., Gade, J., Vahabi, M., Björkman, M., Lavassani, M., Gore, R. N., Lindh, T., & Jiang, X. (2021). Future industrial networks in process automation: Goals, challenges, and future directions. *Applied Sciences*, *11*, 3345.
30. Sufian, A. T., Abdullah, B. M., Ateeq, M., Wah, R., & Clements, D. (2021). Six-gear roadmap towards the smart factory. *Applied Sciences*, *11*, 3568.
31. Tucker, K., Bulim, J., Koch, G., & North, M. M. (2018). Internet industry: A perspective review through internet of things and internet of everything. *International Journal of Management Reviews*, *14*, 26.
32. Bortolini, M., Ferrari, E., Gamberi, M., Pilati, F., & Faccio, M. (2017). Assembly system design in the Industry 4.0 era: A general framework. *IFAC-PapersOnLine*, *50*, 5700–5705.
33. Butt, J., Onimowo, D. A., Gohrabian, M., Sharma, T., & Shirvani, H. (2018). A desktop 3D printer with dual extruders to produce customised electronic circuitry. *Frontiers of Mechanical Engineering*, *13*, 528–534.
34. Gao, J. (2017). Production of multiple material parts using a desktop 3D printer. In *Advances in manufacturing technology XXXI, proceedings of the 15th international conference on manufacturing research, incorporating the 32nd national conference on manufacturing research*, 5–7 September 2017, University of Greenwich, UK. Amsterdam, The Netherlands: IOS Press.
35. Saleh, E., Zhang, F., He, Y., Vaithilingam, J., Fernandez, J. L., Wildman, R., Ashcroft, I., Hague, R., Dickens, P., & Tuck, C. (2017). 3D inkjet printing of electronics using UV conversion. *Advanced Materials Technologies*, *2*, 1700134.
36. Acemoglu, D. & Pascual, R. (2020). Robots and jobs: Evidence from US labor markets. *Journal of political economy*, *128*(6), 2188–2244.
37. Wang, Y., Yuan, L., Ray, Y.Z, & Xun, Xu. (2019). IoT-enabled cloud-based additive manufacturing platform to support rapid product development. *International Journal of Production Research*, *57*(12), 3975–3991.

38. Raju, R., Manikandan, N., Palanisamy, D., Arulkirubakaran, D., Binoj, J. S., Thejasree, P., & Ahilan, C. (2022). A review of challenges and opportunities in additive manufacturing. *Recent Advances in Materials and Modern Manufacturing: Select Proceedings of ICAMMM*, 23–29. Springer.
39. Benkhaira, N, Zouine, N., Fadil, M., Koraichi, S.I., Hachlafi, N.E., Jeddi, M., Lachkar, M., & Fikri-Benbrahim, K. (2023). Application of mixture design for the optimum antibacterial action of chemically-analyzed essential oils and investigation of the antiadhesion ability of their optimal mixtures on 3D printing material. *Bioprinting, 34,* e00299.
40. Wang, Y., Lin, Y., Zhong, R. Y., & Xu, X. (2018). IoT-enabled cloud-based additive manufacturing platform to support rapid product development. *International Journal of Production Research*, 57, 3975–3991.

Chapter 5

Cyber-physical system for advanced manufacturing

Rashmi Nair, Priyanka B. Gaikwad, and Rohit Agrawal

5.1 INTRODUCTION

Advanced manufacturing is undergoing a fundamental shift that ushers in the integration of the physical and digital worlds as enabled by cyber-physical systems (CPSs). CPSs combine with computing, networking, and physical processes and enable real-time interaction between the digital and physical worlds. This is essential to the Industry 4.0 concept – intelligent factories that have interoperable systems, communicate in real time, and are capable of autonomous decision-making (Lee, 2008; Monostori et al., 2016).

CPSs utilize technologies including the Internet of Things (IoT) and sensors, big data analytics, artificial intelligence (AI), and cloud computing to help manufacturers to move to smarter manufacturing processes. While IoT used mainly for the connectivity and data acquisition, AI for the advanced data processing and decision-making, big data analytics for diving deep insights into massive streams of data, and cloud for scalable and flexible infrastructure (Xu et al., 2018). Each of these technologies in turn advances CPS as a whole, enabling production optimization, increased productivity, and improved product quality in manufacturing.

Combining CPSs with manufacturing not only enhances production efficiency but also enables the creation of smart manufacturing ecosystems. Smart manufacturing refers to the ability to create more optimized manufacturing systems through the use of real-time data and adaptive control. Furthermore, CPSs enable predictive maintenance, optimize the supply chain, and promote product lifecycle management, which can lead to innovation and competitiveness in the manufacturing domain (Hehenberger et al., 2016; Jeschke et al., 2017).

DOI: 10.1201/9781003470861-5

5.2 PRINCIPLES AND APPLICATIONS

5.2.1 Principles

The core principles of CPSs include:

Integration: Integration means the computational elements of a system seamlessly interacting with the physical components. This principle assures that computational systems are able to cause and be caused in real time by physical processes. Such integration allows CPS to couple digital control algorithms with physical processes, leading to system performance characterized by increased precision and efficiency (Lee, 2008).

Interoperability: This principle is about the different systems and organizations that should work together efficiently. Interpreting information and making information available so that various systems and components can communicate and cooperate with high-level information is called information fusion (Pivoto et al., 2021).

Real-time Operation: In CPS, real-time operation is very important because the systems need to respond to changes that occur in the physical environment. That is, both input processing and output generation must occur within a pre-specified area of time, which is required for many practical applications (like autonomous vehicles, industrial automation, and medical devices) (Rajkumar et al., 2010).

Feedback Loops: Feedback loops are one of the basic principles of CPS that deals with continuously monitoring and controlling processes that are used to dynamically adjust the system output based on the behavior of the system input and environment. These loops enable the CSP to adapt its activities in response to live data, continuing to provide the best performance and dependability (Baheti & Gill, 2011).

Scalability: This principle states that CPS should maintain robustness over time as data and operations scale across the size of data and parallel operations. The system should be scalable enough so that the system can increase its capacity without changing itself and continue performance and efficiency even if it processes more load and with more operation (Rajkumar et al., 2010).

5.3 CPS INTEGRATION WITH INDUSTRY 4.0 TECHNOLOGIES

Combining Industry 4.0 technologies with CPS is a revolutionary step in the digital manufacturing sector. This convergence optimizes various aspects of the production process, allows it to function better, and makes it more flexible to adapt to be safer.

IoT Integration: IoT is one of the most relevant technologies in the context of CPS because it makes it possible to establish a connection between physical devices and internet services. Sensors and actuators in IoT devices collect the heterogeneous data from the manufacturing environment and then forward the data to CPS for real-time analysis and decision-making, as depicted in CS abstraction. It also allows for the real-time monitoring and optimization of operational efficiency and preventive maintenance (Xu et al, 2018).

Big Data and Analytics: Big data analytics is an enabler for the CPS as it aggregates the huge volume of data that comes off from any source during the manufacturing operations. Manufacturers are able to take this information one step further in the value chain – enabling a better truth on the factory floor by turning to advanced analytics. Predictive maintenance capabilities for predictive analytics can forecast equipment failures, and it, among others, can be used to increase quality control efforts and support further production process optimization that allows cost reduction and higher product quality (Lee et al., 2013).

AI and Machine Learning: The application of complex AI and machine learning algorithms significantly improves CPSs by allowing systems to learn patterns, adapt to novel inputs, and make decisions unassisted. For instance, these technologies are helpful in scheduling, improving throughput, improving process efficiency, performing predictive maintenance, and maintaining the quality parameter with the dynamic process parameter change in real time (Wan et al., 2016).

Cloud Computing: CPS depends on the large-scale data storage, processing power, and analytical capabilities that are offered in the cloud computing framework. With the cloud, CPS can now be made more centralized and can now be integrated and managed at multiple manufacturing sites at scale, thanks to its scalability, flexibility, and access. It leads to better coordination, data synergy, and operational integration (Xu et al., 2018).

Blockchain: It accounts for the integration of highly distributed and secure trustless systems that can have an authoritative and assured trust and integrity over the data within the CPS environment as a decentralized and irrevocable record. It helps to build trust, transparency, and traceability not only throughout the supply chain process but also in other critical manufacturing processes. Tamper-proof nature helps in maintaining data breaches and fraud prevention as well as in ensuring reliable information flows (Bokharaeian et al., 2020).

5.3.1 Applications

CPSs in advanced manufacturing have widespread implications and transformative potentials applied for:

Smart Manufacturing: Continuum modeling develops CPS to improve productivity, adaptability, and quality by integrating real-time integrated data and motion control. Use of sensors and smart machinery integration is also possible in achieving efficient and effective production processes for manufacturers (Baheti & Gill, 2011).

Predictive Maintenance: This application is used to predict equipment failures based on data from the sensors and do predictive maintenance so as to reduce downtime and increase the life of the equipment. Using predictive analytics, manufacturers can predict problems before they cause major downtime (Jeschke et al., 2017).

Supply Chain Optimization: By ensuring real-time tracking and management of supply chain operations, CPS extends real-time visibility and control over the entire supply chain. This involves watching stock amounts, checking shipments, and refining coordination, thus requesting and cutting expenses (Wang et al., 2016).

Product Lifecycle Management: The data from design and manufacturing processes, services, and disposal are integrated across the complete product lifecycle at the CPS. This type of holistic view allows for better decisions, better products, less waste in production, and better post-sales services (Hehenberger et al., 2016).

5.4 CONTEXT-AWARE COMPUTING FOR CPS

It can be defined as the ability of the system to be aware of, understand, and adapt to the contextual information, characteristics of the environment, and usage. Context-aware computing additionally supports a CPS system to make good decisions and to perform its tasks well (Chen & Kotz, 2000).

5.4.1 Applications in manufacturing industry

By incorporating availability of context into the development of CPSs, it is possible for manufacturers to attain flexible and efficient manufacturing operations with improved safety, resulting in a better performance of the overall manufacturing process (Broy et al., 2012).

Dynamic Resource Allocation: Context-aware CPS could monitor and analyze real-time data on demand and supply conditions. For instance, if the order volume suddenly experiences a surge, the CPS can assign the necessary resources like raw materials, machinery, and manpower to ensure that the sudden increase in demand is met. When the demand decreases, the system can reduce resource use to avoid overproduction and waste. While following, the production schedules can be increased, and while cutting back, there is no pressure to work overtime, which helps the company to reduce operational costs altogether.

Adaptive Control Systems: As the name would imply, adaptive control systems adjust how the machine runs based on different changing conditions (e.g., temperature, humidity, or how worn down the equipment is). Information gathered from sensors placed throughout a company's production environment, for instance, can predict when and how different variables might impact production quality or the operation of machinery. Adjusting to these changes will allow the CPS to receive production in ideal conditions, maintain the quality of the product, and extend the life of its machinery.

Human-Machine Interaction: The most critical part of human-machine interaction within a context-aware CPS is the design of better interaction interfaces and workflows that could satisfy human operators demands and conditions. These systems can alter the interface and controls to best meet the needs of the builder based upon the builder's experience, fatigue, or other elements of workload. Additional assistance might come in the form of simplifying the interface for a new worker or helping when an operator is stressed or fatigued. This raises safety and minimizes the chance of human error while boosting operational efficiency as well since operators are able to work more effectively.

5.5 CYBER-PHYSICAL SYSTEMS IN INDUSTRIAL ROBOTICS

The integration of CPSs in industrial robotics is an innovation that works in a transformative manner for industrial manufacturing, combining a higher level of automation, efficiency, and safety.

More Automation and Flexibility: This is also one of the biggest gains that industrial robotics will achieve with CPS. CPSs are an essential tool for data acquisition and decision-making, making the implementation of real-time data exchange and adaptive control possible, leading to a system with high flexibility, even allowing robots to perform more complex and precise tasks.

This enables the creation of co-robots, or cobots, where humans and robots collaborate and work side by side, making it safe for the human and providing more production (Monostori et al., 2016).

Autonomous Decision-Making: As industrial robots with CPS are based on real-time data and advanced algorithms, they make decisions based on preset conditions without human interference. In applications such as automated quality inspection, adaptive assembly lines, and dynamic scheduling, robots must adapt promptly and accurately to changing conditions, and to this end, real-time learning is paramount (Wan et al.

Human-Robot Collaboration: With the degree of advanced sensors, artificial intelligence, and real-time monitoring systems integrated in CPS, human-robot collaboration is no longer possible to make human-robot collaboration safer and more efficient. These technologies will guarantee that robots can handle the human environment in order to solve environmental changes or to solve varying problem capacities in real time (Villani et al., 2018).

5.6 GREEN CPS

Green CPS is contributing to the development of environmentally friendly and energy-efficient industrial processes. Green CPS makes use of cutting-edge technologies in order to reduce environmental footprints and enhance resource efficiency and also to develop sustainability in various verticals.

5.6.1 Energy efficiency

Energy efficiency is a vital requirement for green CPS. A range of mechanisms are used in these systems to reduce energy usage, such as:

Live Energy Monitoring: Tracking energy for your day-to-day use measures improvements that drive energy usage and cuts it down immediately.

Predictive Maintenance (Green CPS): Green CPS can also improve the overall manufacturing process by minimizing machine downtime and unnecessary energy consumption that relates to older or not up-to-scratch machinery by processing data to predict when the maintenance is needed.

Adaptive Control Systems: The adaptive control systems monitor the demand for real time and manipulate the machine operation to enhance its efficiency, without sacrificing the performance (Wu et al., 2016).

5.6.2 Waste reduction

Since the green CPSs produce less waste, the waste reduction is one of the emphatic points in which the green CPSs are better. Through data analytics and real-time monitoring, Lettiere says they have the capability of identifying and reducing waste all over the shop floor in manufacturing processes. It thus encompasses the minimization of materials waste, the improvement of product quality, and closed-loop recycling schemes for well-recycling of materials (Zhou et al., 2016).

5.6.3 Sustainable supply chains

Green CPSs improve transparency and traceability to generate sustainable supply chains. IoT and blockchain are amongst the technologies that are proving vital when it comes to how products are sourced, made, and transported in a sustainable way. Green CPS, by providing the means to trace the origin and path of the products, are capable of promoting responsible consumption and production practices and increasing long-term sustainability as well (Pagoropoulos et al., 2017).

5.7 BARRIERS TO ADOPTION OF CPS

Apart from the several advantages that CPS brings with it, there are also widespread barriers keeping the CPS uptake at bay in manufacturing. There are several technological barriers to the deployment of CPS, including complexities integrating CPS with existing legacy systems and interoperability with devices and platforms (Kang et al., 2016). Another key challenge is financial restrictions, a major barrier – especially for small- and medium-sized enterprises (SMEs) – that often have difficulty in meeting the high costs of initial investment needed for CPS implementation (Lee et al., 2015).

Resistance from within organizations and cultures also plays a part in halting CPS implementations. Employees are usually slow to embrace new technologies because they might not know how they work or they fear that these robots would take over their jobs. This concern can be addressed through effective change management strategies and continuous training in order to enable employees to work with these new technologies (He & Xu, 2015). As more devices become connected, data security and privacy are equally important issues to solve since more connectivity leads to information exposure to cyber threats. Proper cybersecurity protection is a must but can be a costly and complicated matter (Jeschke et al., 2017).

CPS adoption is also hampered by regulatory and compliance hurdles. Navigating the different industry standards and regulations can be a lot of work, which is especially true in sectors that are highly regulated (Kang et al., 2016).

Technological Barriers: Technological barriers include the complexity of interfacing CPS with the existing legacy systems, interoperability across different devices and platforms, and real-time performance (Kang et al., 2016).

Financial Constraints: Enormous financial outlay is required to implement CPS mainly for new equipment, rejuvenation of existing facilities, employee training (Lee et al., 2015) which is particularly problematic for SMEs.

Organizational and Cultural Resentment: Any change within an organization is susceptible to criticism, since adapting to new technologies is not an easy task, especially for employees who are used to their routines. He and Xu (2015) indicated that the impendence can be caused by the lack of understanding, job loss, or the changing of the work process.

Cybersecurity: As these CPS have become more connected, they are becoming increasingly vulnerable to cyberattacks, data breeches, and security threats. The problem is the nature of security and privacy awareness; data security and privacy are most important; however, implementing robust cybersecurity countermeasures is complicated and expensive (Jeschke et al., 2017).

Regulatory and Compliance: Compliance with various industry standards and regulations can pose a huge challenge, especially in heavily regulated industries like aerospace and healthcare. It will take a lot of work and a deep understanding to negotiate these regulations (Kang et al., 2016).

5.8 FUTURE TRENDS OF CPS IN INDUSTRY 4.0

CPSs are used in many areas of advanced manufacturing. In addition to helping with automation and production flexibility, they help in the implementation of predictive maintenance and real-time supply chain optimization (Hehenberger et al.). CPSs enable human-robot collaboration, for example, in industrial robotics, where CPSs make it possible for humans and robots to work together in terms of safety and efficiency, increasing production and safety in manufacturing environments (Villani et al., 2018).

Emerging trends with respect to sustainability in manufacturing have given birth to a new category of CPS called green CPS. Large irrigation projects are designed to use and operate (Zhou et al., 2016) systems, which enables better integration, leading to energy conservation, waste reduction, and sustainable supply chains. Energy saving: Green CPS facilitates real-time energy monitoring and adaptive control systems, focusing on more sustainable manufacturing practices (Wu et al., 2016).

However, the future of smart manufacturing relies on the constant advancement of CPS technologies. Further research should formulate a stronger packaged assessment approach of CPS readiness and investigate the combination of emerging technologies like 5G, AR, and blockchain with CPS (Xu et al., 2018). A success in the face of some difficult challenges for CPS in advanced manufacturing requires a synergistic effort from academia, industry, and policymakers.

5.9 DISCUSSION

CPSs have revolutionized and promoted the technological progress of manufacturing. Integration of CPS into Industry 4.0 technologies (e.g., IoT, AI, and big data analytics) leads toward smart factory, where machineries, systems, and humans are connected with each other so that interaction takes place in real time (Kagermann et al., 2013). The IoT gathers information and sends it to a CPS that provides additional limitations on manufactural procedures permitting real-time monitoring and control. This connectivity improves manufacturing productivity, reduces downtime, and supports predictive maintenance (Xuelsr et al., 2018).

Further, CPS with AI and machine learning algorithms provides more advanced data processing and decision-making capabilities. Such technologies facilitate adaptive and autonomous decision-making for better production processes and quality (Wan et al., 2016). Big data analytics are empowered to assist CPS with processing huge amounts of data created during manufacturing operations for the sake of offering good manufacturing process optimization and quality control through the collection of real-time information (Lee et al., 2013).

CPS also includes the role of cloud computing on which CPS highly depends on for data storage, processing, and analysis; community clouds play a crucial role. It enables the scalability and flexibility needed to effectively interoperate CPS on varied manufacturing sites (Xu et al., 2018). On the other side, CPS can benefit from blockchain technology by leveraging guaranteed data security and integrity, offering decentralized and tamper-proof records, which are necessary for preserving trust and transparency throughout the manufacturing processes (Bokharaeian et al., 2020).

5.10 CONCLUSION

CPSs are a game changer for digitalized production in providing efficiency, flexibility, and resource-saving. To address the challenges that accompany even small-scale use, a mix of technological, organizational, workforce, and regulatory solutions are necessary, and the United States has made progress in some of these areas over the last decade. Research and collaboration will be needed moving forward in order to take full advantage of what CPSs can do for advanced manufacturing as the digital transformation progresses.

REFERENCES

Baheti, R., & Gill, H. (2011). Cyber-physical systems. *The Impact of Control Technology*, 12(1), 161–166.

Bokharaeian, B., Badawi, S., & Bergmann, R. (2020). Cyber-physical system meets blockchain: A decentralized and trusted design and deployment concept. In *International Conference on Industrial Informatics (INDIN)* (pp. 761–768). IEEE.

Broy, M., Cengarle, M. V., & Geisberger, E. (2012). Cyber-physical systems: Imminent challenges. In *Proceedings of the 37th International Conference on Current Trends in Theory and Practice of Computer Science* (pp. 1–6). Springer. https://doi.org/10.1007/978-3-642-34059-8_1

Chen, G., & Kotz, D. (2000). *A Survey of Context-Aware Mobile Computing Research*. Department of Computer Science, Dartmouth College.

He, W., & Xu, L. D. (2015). Integration of distributed enterprise applications: A survey. *IEEE Transactions on Industrial Informatics*, 10(1), 35–42. https://doi.org/10.1109/TII.2013.2295035

Hehenberger, P., Vogel-Heuser, B., Bradley, D., Eynard, B., Tomiyama, T., & Achiche, S. (2016). Design, modelling, simulation and integration of cyber-physical systems: Methods and applications. *Computers in Industry*, 82, 273–289. https://doi.org/10.1016/j.compind.2016.05.006

Jeschke, S., Brecher, C., Song, H., & Rawat, D. B. (2017). *Industrial Internet of Things: Cybermanufacturing Systems*. Springer.

Kagermann, H., Wahlster, W., & Helbig, J. (2013). *Recommendations for Implementing the Strategic Initiative INDUSTRIE 4.0. Securing the Future of German Manufacturing Industry. Final Report of the Industrie 4.0 Working Group*. Forschungsunion.

Kang, H. S., Lee, J. Y., Choi, S., Kim, H., Park, J. H., Son, J. Y., ... & Noh, S. D. (2016). Smart manufacturing: Past research, present findings, and future directions. *International Journal of Precision Engineering and Manufacturing-Green Technology*, 3(1), 111–128. https://doi.org/10.1007/s40684-016-0015-5

Lee, E. A. (2008). Cyber-physical systems: Design challenges. In *2008 11th IEEE International Symposium on Object and Component-Oriented Real-Time Distributed Computing (ISORC)* (pp. 363–369). IEEE.

Lee, J., Bagheri, B., & Kao, H. A. (2015). A cyber-physical systems architecture for Industry 4.0-based manufacturing systems. *Manufacturing Letters*, 3, 18–23. https://doi.org/10.1016/j.mfglet.2014.12.001

Lee, J., Kao, H. A., & Yang, S. (2013). Service innovation and smart analytics for Industry 4.0 and big data environment. *Procedia CIRP*, 16, 3–8. https://doi.org/10.1016/j.procir.2013.02.001

Monostori, L., Kádár, B., Bauernhansl, T., Kondoh, S., Kumara, S. R. T., Reinhart, G., ... & Ueda, K. (2016). Cyber-physical systems in manufacturing. *CIRP Annals*, 65(2), 621–641. https://doi.org/10.1016/j.cirp.2016.06.005

Pagoropoulos, A., Pigosso, D. C., & McAloone, T. C. (2017). The emergent role of digital technologies in the circular economy: A review. *Procedia CIRP*, 64, 19–24. https://doi.org/10.1016/j.procir.2017.03.178

Pivoto, D. G., De Almeida, L. F., da Rosa Righi, R., Rodrigues, J. J., Lugli, A. B., & Alberti, A. M. (2021). Cyber-physical systems architectures for industrial internet of things applications in Industry 4.0: A literature review. *Journal of manufacturing systems*, 58, 176–192.

Rajkumar, R., Lee, I., Sha, L., & Stankovic, J. (2010). *Cyber-physical systems: The next computing revolution*. In *Design Automation Conference* (pp. 731–736). IEEE.5.1.3 Applications. https://doi.org/10.1145/1837274.1837461

Villani, V., Pini, F., Leali, F., & Secchi, C. (2018). Survey on human-robot collaboration in industrial settings: Safety, intuitive interfaces and applications. *Mechatronics*, 55, 248–266. https://doi.org/10.1016/j.mechatronics.2018.02.009

Wan, J., Zhang, D., Sun, Y., Lin, K., & Yang, L. T. (2016). Toward dynamic resources management for IoT-based manufacturing. *IEEE Communications Magazine*, 54(6), 68–74. https://doi.org/10.1109/MCOM.2016.7481383

Wang, S., Wan, J., Zhang, D., Li, D., & Zhang, C. (2016). Towards smart factory for Industry 4.0: A self-organized multi-agent system with big data-based feedback and coordination. *Computer Networks*, 101, 158–168. https://doi.org/10.1016/j.comnet.2015.12.017

Wu, Q., Lin, Y., & Guo, X. (2016). Green CPS: A new CPS design paradigm for sustainable and energy-efficient Internet of Things. In *2016 IEEE International Conference on Services Computing (SCC)* (pp. 399–406). IEEE.

Xu, L. D., Xu, E. L., & Li, L. (2018). Industry 4.0: State of the art and future trends. *International Journal of Production Research*, 56(8), 2941–2962. https://doi.org/10.1080/00207543.2018.1444806

Zhou, K., Liu, T., & Zhou, L. (2016). Industry 4.0: Towards future industrial opportunities and challenges. In *2015 12th International Conference on Fuzzy Systems and Knowledge Discovery (FSKD)* (pp. 2147–2152). IEEE. https://doi.org/10.1109/FSKD.2015.7382284

Chapter 6

Digital twins for advanced manufacturing

Priya Subha M., Sathish V., and Janardhan Vistapalli

6.1 INTRODUCTION

Monitoring of system status is a major task in most of the industries to predict or prevent the system failure and enhance efficiency. Engineers and operators used online monitoring systems to spot anomalies and patterns and decide on preventive maintenance (Ge, 2008). Real-time monitoring keeps track of the application status continuously, enabling an immediate response. Online and real-time monitoring are equally important, although they have different features and areas of emphasis. Real-time monitoring prioritizes important occurrences, while online monitoring guarantees ongoing data access (Feng, 2021). Digital twin (DT) is one of the real-time monitoring technologies that gives a feel of how the actual system is behaving at present by creating a virtual replica of the system and its movements with sensor data. Detailed information related to DT is discussed in subsequent sections.

6.1.1 Definition of digital twin

DT is defined as "an integrated multiphysics, multiscale simulation of a vehicle or system that uses the best available physical models, sensor updates, fleet history, etc., to mirror the life of its corresponding flying twin" by the American Society of Mechanical Engineers (ASME) (Rosen, 2015).

Similar to the DT, the term "product avatar" was given by Hribernik et al. (Hribernik, 2006). The terms like "Digital model" and "Digital shadow" are also introduced by researchers (Kritzinger, 2018). Many organizations and researchers gave their own interpretations of DT (Ríos, 2015; Madni, 2019; evtić, 2023; Ghorbani, 2024). The evolution of the DT definition is described in Figure 6.1.

DOI: 10.1201/9781003470861-6

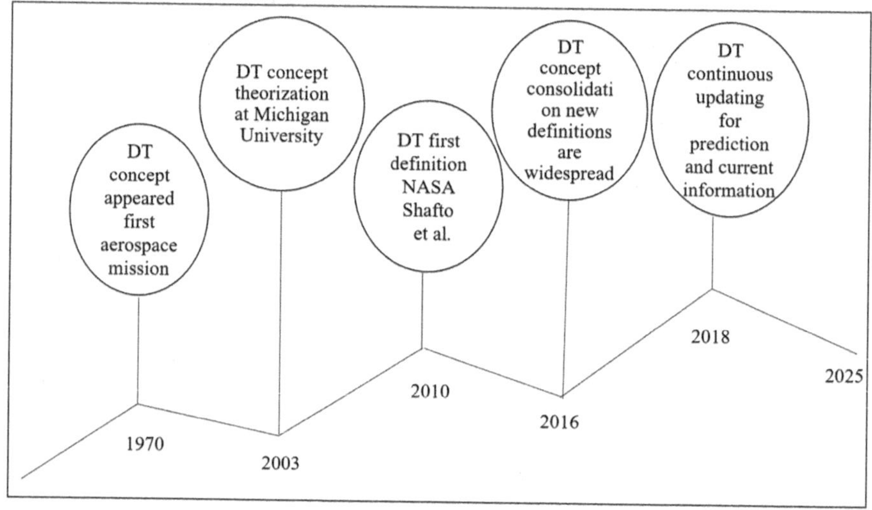

Figure 6.1 Definition of digital twin (Agnusdei, 2021).

6.1.2 Applications of digital twin technology

DT gained attention across various industries due to its advanced features, such as real-time monitoring and simulation (Lu, 2020), predictive maintenance, optimization and decision support (Tao, 2018), visual testing and prototyping, and real-time monitoring and control (Glaessgen, 2012). Due to this potential spread, DT spread to diverse sectors like agriculture, construction, automobiles, and healthcare, as described in Figure 6.2. The demand for DT surged during the Covid-19 pandemic (Singh, 2021). Applications of DTs in various fields are described below.

6.1.2.1 Manufacturing industry

DT is used in the fabrication of complicated devices for design interaction prior to fabrication (Yao, 2023). The DT revolutionized model-based systems engineering (MBSE), offering a view of a system's entire lifecycle (Singh, 2021). DT promotes the monitoring and analysis of product quality, allowing for early defects identification and deviations from specifications (Jayal, 2010). DT provides visibility into supply chain operations by modeling inventory, transportation, and production processes (Kritzinger, 2018).

6.1.2.2 Healthcare sector

DT is used for predictive medical device maintenance, performance optimization, and hospital lifecycle optimization (Barricelli, 2019). GE Healthcare

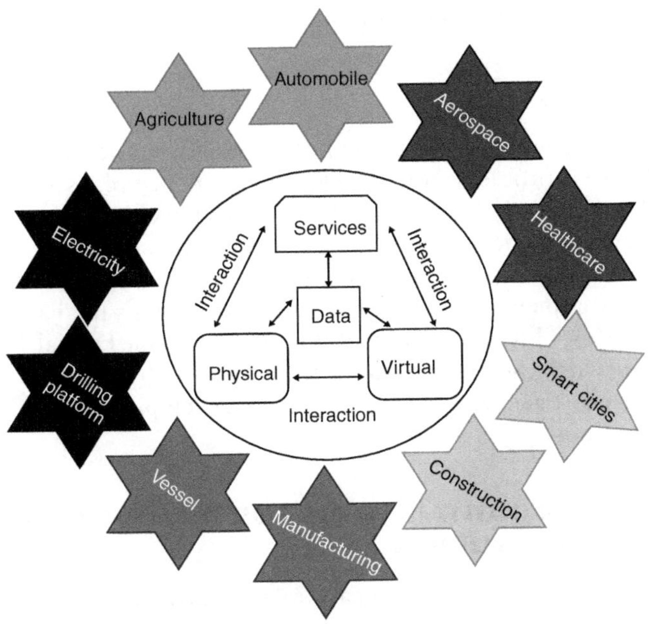

Figure 6.2 Application of digital twin technology (Qi, 2021).

formed a "Capacity command centre" in the Johns Hopkins Hospital in Baltimore to improve decision-making. Siemens Healthineers created a DT to optimize the Mater Private Hospitals (MPH) in Dublin (Barricelli, 2019; Tao, 2019). Dassault launched the Living Heart Project (LHP) to create a digital heart that can replicate the functions of an actual heart (Aliper, 2016). Liu et al. offered a unique, universal, and scalable architecture that can track, identify, and forecast every facet of a person's health. DTs are used to create personalized models of the patients by incorporating physiological data, medical imaging, and genetic information (Miotto, 2017). DT provides remote monitoring of a patient's health state for telemedicine services (Patel, 2012).

6.1.2.3 Agriculture industry

DT is used for monitoring dairy cattle remotely to detect health issues for smart farming systems (Tao, 2019). DT could also be used to measure and learn about the composition, capacity, and quality of the soil and crop for irrigation management (Singh, 2021). IBM's Watson Decision Platform and John Deere's Digital Farming solutions leverage DT to help farmers and optimize farm operations and increase productivity. DT is used to provide

AI-driven insights for dairy farmers, helping them optimize herd management and increase milk production.

6.1.2.4 Aerospace

NASA has constructed a DT to replicate the real-world circumstances in order to forecast problems and develop maintenance plans for an aircraft in orbit (Tuegel, 2011). Currently, NASA also makes an effort to employ the DT to ensure crew safety in the event that the aircraft is not in operation. In an effort to save costs and increase production efficiency, Northrop Grumman has implemented the DT on the F-35 fighter's final assembly line, much like NASA. (Tao, 2019; Yao, 2023). The Airframe Digital Twin (ADT) is a DT framework used in aeronautical engineering (Hu, 2021). Australia's national science agency, CSIRO, has been developing DT solutions for mining operations.

6.2 REVIEW ON DIGITAL TWINS FOR VARIOUS ASPECTS OF MANUFACTURING

Some major aspects of DT, which captured the attention of manufacturing industries and researchers, include predictive maintenance, simulation and training, energy efficiency, and DT-assisted services. A review of research status of the aforementioned aspects is given in the following subsections.

6.2.1 Predictive maintenance

The meaning of "Predictive maintenance" can be understood from its name itself. A sophisticated predictive maintenance technique, which seeks to predict the breakage of the machinery or equipment, is done by Chen (2017). Many used DT for predictive maintenance of machines; Liu (2019) used it for aero engine bearings; Qiao (2019) for milling machine tools. DTs are used for fault diagnosis and remaining usable life estimate. Fault detection (Zhong, 2023) is achieved by using a supervised learning technique, namely a support vector machine (SVM). Aivaliotis et al. (You, 2022) focused his research on gearbox bearing using "The remaining usable life" (RUL) to provide accurate results.

6.2.2 Simulation and training

Simulation and training are essential methods for improving efficiency, streamlining the procedures, and reducing risk in the field of DT technology. DT also helps in the improvement of human-machine interaction (Wilhelm, 2021). Virtual simulation technology's purpose was not to perform analyses or optimization tasks on tangible objects but rather to create immersive experiences and simulated environments. Virtual simulation technology relied on intricate models and data to replicate the hand; DT leveraged their sensing, diagnostic, and forecasting capabilities to observe and refine the real-time states of physical entities (Yao, 2023). Using DT, integration of the real world into virtual reality (VR) is done for VR-based teaching and learning on a mobile platform (Yao, 2023).

6.2.3 Energy efficiency

In the industry, DT assists in the monitoring and optimization of energy use. Researchers provide methods and architecture for analyzing energy consumption on a machine process level (Mawson, 2019). Cloud-based DT is used in the monitoring system for wind farm development and monitoring (Fuller, 2020). DT is also used for forecasting a wind turbine's health (Yao, 2023).

6.2.4 Digital twin for service

DT is majorly popular due to the services it offers. The DT-based services are monitoring, evaluation, simulation, prediction, optimization, and control. DT-based monitoring service is mainly useful for monitoring geometrical models, finite element models, physical element behavioral changes, and operating status (Liu, 2022). The simulation service provides customers with virtual tests, design verification, and virtual communication (Meierhofer, 2020). The prediction services bring the prediction of the future state of the systems into the present reality (Zhong, 2023). The optimization services can serve as a trial-and-error and decision-making process from the physical world to the digital world. The control services use the feedback from the decision of the optimization service to control the physical world so that it can adapt to the changes in the environment (Wang, 2021; Tao, 2022). A description of different elements that make a DT is briefed in the following subsection.

6.3 ELEMENTS OF A DIGITAL TWIN

The DT has many elements and tools that need exclusive focus for implementation in the real world. These include things like physical and virtual parts, data acquisition, data synchronization, IoT, and CPS.

6.3.1 Fundamental elements of digital twin

Envisioned by the visionary Grieves, there are three fundamental elements, which are given here (Grieves, 2014).

6.3.1.1 *Physical part*

The object or system in the physical world is named as the physical part (Tao, 2018). It may be a building, an industrial robot, a wind turbine, a piece of manufacturing equipment, or even the whole infrastructure of a city (Madni, 2019; Zhu, 2019). This part represents the material composition, shape, physical attributes, and functional behavior of the real item. Sensors, cameras, and other monitoring devices placed on or close to the actual object are used to gather the information of the physical part (Redelinghuys, 2020).

6.3.1.2 *Digital part*

The virtual equivalent of the physical part is the digital part (Junnan, 2022). It exists in the digital realm and mirrors the real-world object. The 3D CAD models and additional design representation are included in the DT. It depicts the physical asset's size, form, and spatial connections. These models mimic the behavior of the real asset in various scenarios (Pang, 2023). It considers elements like axis and restriction. Data from several sources, such as historical documents (Yao, 2023), maintenance logs, sensors, and IoT are integrated into the digital part. The digital part evolves throughout the asset lifecycle, from design to maintenance.

6.3.1.3 *The connection*

The connection or link creates a bridge between the digital and physical parts. Data is transferred between the physical part and the digital counterpart through wired or wireless networks. These gadgets gather data from the physical asset in real time (Zhu, 2019) and ensure that the changes are being implemented on the digital part. DT hosted by cloud services provides scalability, accessibility, and safe storage. Once the connection between real and digital is complete, DTs are further integrated with Dashboards, 3D Visualizations, Augmented Reality (AR) interface, etc. Due to the integration of all these tools, the DT technology, which was three-dimensional is transformed to five dimensions, where services (Sun, 2020) and data are the new dimensions (Tao, 2022).

The service system covers alarm management, process control, problem diagnosis, prognosis, and maintenance (Qiao, 2019). "Data" include design parameters, process data, and feedback data. One of the key components to collect data from physical and virtual aspects is data acquisition, which can provide accurate and comprehensive information. To integrate the collected data with the digital part, IoT is a key tool (Tao, 2019).

6.3.2 Data acquisition and IoT

Data acquisition is the process of collecting, measuring, and digitizing real-world data from various sources such as sensors and instruments for further processing and analysis. It involves capturing analog or digital data from physical systems and converting them into digital form for the computer to process. Data acquisition forms a backbone of interconnected things. The acquired data is then transmitted over networks, often wirelessly, to central servers or cloud platforms for processing and analysis. IoT extends the capabilities of data acquisition by enabling the seamless integration of diverse devices and sensors into a unified ecosystem. By leveraging data acquisition techniques, IoT systems can collect and analyze vast amounts of data from distributed sensors, enabling intelligent decision-making, predictive maintenance, real-time monitoring, and optimization of processes and services. Due to the integration of IoT technologies, AI, and ML into DT, there arose a demand for creating and integrating DT for multiple components or system of systems. Due to this, the physical part and digital part of a DT became the physical world and digital world, which are explained below.

6.3.3 Physical and virtual world

"Physical world" stands for the actual physical infrastructure, which is made up of several physical parts or systems related to an industry or production system. All the equipment and processes related to raw materials, parts, tools, fixtures, and infrastructure (Negri, 2017; Zhu, 2019) as well as environmental factors that affect worker safety, product quality, and production efficiency collectively form the physical world (Plessis, 2016).

"Virtual world" is a collection of various virtual parts, which are untouchable. All the models that exist in the virtual world represent a digital counterpart or replica of the physical world (Zhu, 2019). Using computer-aided design or computer-aided engineering software, virtual models are produced, which are digital replicates of the actual assets and processes (Wynn, 2023).

As the physical and the virtual world are growing bigger, there is a need for synchronization between these two worlds to improve accuracy and efficiency. This requires proper synchronization of the data that are collected from both physical and virtual worlds, details of which are discussed in the following subsection.

6.3.4 Data synchronization and real-time updates

"Synchronization" means to be in line with each other. For DT, the data from both physical and virtual worlds have to be one-to-one for accurate functioning. To achieve this, DT uses many synchronization methods or techniques for exchanging information and providing real-time updates (Ghorbani, 2024). Some methods are given, such as:

- *Time-Based Synchronization:* Systems synchronize based on time signals or clocks.
- *Event-Based Synchronization:* Synchronization occurs in response to specific events or triggers.
- *Hardware-Based Synchronization:* This can include hardware interrupts, synchronization primitives provided by processors, or dedicated hardware modules designed for synchronization tasks.
- *Software-Based Synchronization:* Synchronization is managed through software mechanisms such as locks, semaphores, mutexes, or condition variables.
- *Phase-Locked Loop (PLL):* PLLs are generally used for clock synchronization in digital systems, ensuring that local clocks remain aligned with external reference clocks.

Instantaneous updates along with data synchronization help to keep an eye on their manufacturing procedures, identify irregularities, and take immediate remedial action to boost productivity, quality of output, and customer happiness (Capra, 2014).

6.3.5 Cyber-physical systems vs digital twin

CPS is a system that integrates computation, networking, and sensing into the physical objects and processes (Gunes, 2014). CPS is essentially an embedded system in which computers are incorporated into a mechanical or electrical system to perform specific, focused activities within the constraints of real-time computing (Monostori, 2016). The integration of computing capabilities with physical systems enables precise control and monitoring, revolutionizing various industries. DT is an advanced version of CPS. CPS integrated with a digital world resembles a DT. The following section explains the framework and tools required for realizing a DT.

6.4 FRAMEWORKS FOR DEVELOPING A DIGITAL TWIN

Framework is a carefully crafted masterpiece comprising several intricate stages to develop a technology (Zheng, 2019; Shao, 2020; enaLi, 2021). There are several stages involved in developing a DT for manufacturing systems; all these form the DT framework. The several stages of the DT framework are:

- Identifying the key objectives of making DT.
- Modeling and simulation.

- Data acquisition and integration.
- IoT integration and connectivity.
- Data visualization and human-machine interface design.
- Deployment and testing.
- Feedback and optimization.

The first stage of developing DT for any case is to decide the targets and scope. It may be improving the performance, real-time monitoring, energy efficiency, or offering any service of a specific machine or manufacturing system. Once the objective and scope are decided, the approach for creating a digital replica is described in the following subsection.

6.4.1 Making digital replica

Detailed digital reproductions are produced using a variety of methods; these include 3D scanning, simulation software, computer-aided design (CAD) (Myers, 1998; Chryssolouris, 2009; Fang, 2021), computer-aided manufacturing (CAM), computer-aided process planning (CAPP), manufacturing resource planning (MRP), etc. (Farsi, 2020; Warke, 2021). The amount of information needed, the system's complexity, and the accessibility of data or model validation and calibration are all the important factors to be considered for choosing an appropriate method.

CAD models can capture geometric dimensions (Hu, 2021), material properties, and other design specifications necessary for accurate representation. Simulation software provides basic physics-based models to sophisticated computational fluid dynamics (CFD) and finite element analysis (FEA) capabilities. Each field has its platforms and tools, such as ANSYS's Twin builder, Microsoft's Azure and MATLAB Simulink (Singh, 2021; Yao, 2023), and ADAMS for doing multibody dynamic analysis, etc. DT models are interactive in terms of data feedback and control. They are helpful for improving the conventional product design and development process, whereas the CAD model is used to do only simulation.

Once the digital replica of the physical system is created, to interface the physical system with the digital replica, a multitude of data has to be obtained from the physical part. The physical system has to be integrated with many sensors for this.

6.4.2 Sensor integration

The real-time data gathering from physical assets and processes is crucial for making a DT; sensor integration is crucial for this. The behavior and function of a system are better understood by utilizing several parameters that sensors record; these include temperature, pressure, vibration, flow rates, etc. Sensor types that are suitable for the production environment must be chosen, their placement must be planned, and data-transmission connection protocols must be set up before sensors can be integrated into DT. The following paragraphs describe a brief about these parameters.

6.4.2.1 Sensor types

Many types of sensors are used in industrial environments, based on the operation, application, measured qualities, and technologies. These are further classified into various types depending on their parameters as described in Figure 6.3.

Based on the operation, sensors are classified as active and passive sensors. The active sensors are those that can function and generate an output signal that changes depending on the measured quantity; these need an external power source. Passive sensors produce output signals in response to external stimuli and do not need any external power source (Veverka, 2020).

Based on measured quantities, sensors are classified into various types, such as temperature sensors, pressure sensors, humidity, light, and proximity.

The sensors, which are based on the technology, are classified as inductance sensors, resistance sensors, capacitance sensors, etc. (Harpe, 2021; Moheimani, 2022).

One needs to choose a particular sensor for making a DT based on their objective or application. Some of the sensors, which are often used for DT in applications, are temperature sensors, pressure sensors, accelerometers, gyroscopes, encoders, potentiometers, force sensors, flow meters, level sensors, proximity sensors, position sensors, gas sensors, and humidity sensors.

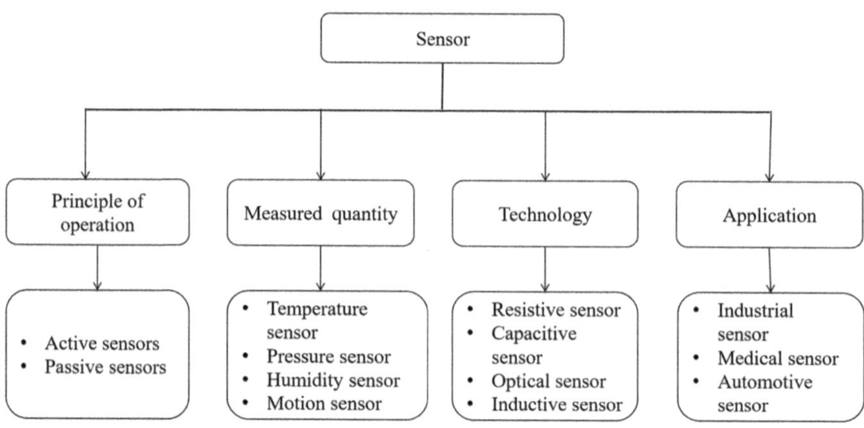

Figure 6.3 Classification of sensors.

6.4.2.2 *Sensor deployment*

Sensor deployment is an important problem that can affect various operations. Key points to be considered in sensor deployment for continuous and seamless data acquisition are mounting location (Iwata, 2017), sensor range and reach (Zou, 2004), minimizing blind spots and gaps in data collection and mitigating data loss. A robust communication network reduces the effects of disturbance or breakdowns in the network, guaranteeing continuous data transfer to the DT platform.

6.4.3 Data management

Data management is an important aspect of DT; it involves data acquisition, data storage, and data processing (Hu, 2021). All these stages are crucial to developing an efficient and successful DT.

6.4.3.1 *Data acquisition*

Data acquisition is the act of gathering information from different sensors and devices (Uhlemann, 2017b; Liu, 2023). Tools for data acquisition in creating a DT vary depending on the nature of the object or system being modeled and the specific parameters that need to be measured. Here are some common tools and techniques used for data acquisition:

Data Loggers: Data loggers are standalone devices or modules that record data from sensors over time.

DAQ (Data Acquisition) Systems: They consist of hardware and software components designed to capture, process, and analyze signals from various sensors and sources. DAQ systems are highly customizable and can be tailored to specific data acquisition requirements. Various types of boards extensively used for data acquisition purposes are National Instruments DAQ boards, Arduino boards with compatible shields, Raspberry Pi boards with additional modules, TI LaunchPad development kits, BeagleBone Black boards with expansion options, and LabJack data acquisition devices. In addition to the classical approaches or tools mentioned before, researchers developed techniques for data acquisition using the internet and AI (Uhlemann, 2017a; Zhu, 2019). Some of the data acquisition tools that depend on the internet are IoT devices and cloud platforms.

IoT devices: These are physical objects embedded with sensors, actuators, and other technologies that connect and use the internet to communicate with other systems and devices. These devices frequently use wireless communication protocols, including Bluetooth,

Zigbee, Lora WAN, or Wi-Fi, to send data to the centralized servers for analysis and storage. IoT devices are ideal for DT applications because they allow for the real-time monitoring and management of physical assets and processes. IoT relies heavily on radio-frequency identification technology (Farsi, 2020). IoT technologies like bar codes, QR codes, and RFID store specific information (Hu, 2021).

Cloud platforms: Scalable services for gathering, storing, and analyzing data from dispersed sensors are offered by cloud-based data collection platforms. With features like data intake, storage, analytics, visualization, and ML capabilities, these platforms are ideal for creating and implementing DT on a large scale. Some of the platforms used for cloud platform interfacing are AWS IoT, Azure IoT, Google Cloud, IBM, Firebase, and ThingSpeak.

6.4.3.2 Data storage

The second process of the data management is to store the data. Data storage is the process of keeping the gathered information for later management, processing, and analysis (Qi, 2021). Large volumes of data produced by sensors and other data sources must be stored for further analysis and display in a data repository or database (Lu, 2020) that the DT can access (Barricelli, 2019). To do this, data from dispersed sensors may need to be combined, using edge computing or gateway devices (Cimino, 2019), before being sent to the cloud or onsite servers. Some examples of data storage options are conventional relational databases, NoSQL databases, and time series databases, which are designed to manage time-stamped sensor data (Hu, 2021; Qi, 2021). Some of the cloud-based database platforms are mentioned in subsection 6.4.3. Cloud storage has become a well-liked alternative to conventional storage systems for managing and storing data. To maximize storage performance and efficiency, data storage systems usually use the hierarchical storage management (HSM) approach.

6.4.3.3 Data processing

Data processing is the manipulation and transformation of unprocessed data into insights or useful information using a variety of computational and analytical methods. The data processing module is capable of creating a live representation of a physical item by extracting information from heterogeneous multisource data to assist decision-making, problem-solving, and business operations across a variety of sectors and disciplines (Lu, 2020). The commonly used data processing technique is cloud-edge computing (Qi, 2021). The stages of data processing include pre-processing, filtering, sorting, grouping, summarizing, joining, and modeling (Iwata, 1995; Shafto, 2010; Bao, 2018). Once the data are processed, different communication protocols are used to get the suitable data to the digital world.

6.4.3.4 *Communication protocols*

Data transmission from sensors to the DT platform requires the use of communication protocols. Many communication protocols, such as MQTT, HTTP, CoAP, and OPC UA, are available (Asavasirikulkij, 2022). A popular publish-subscribe messaging protocol for the IoT is MQTT (Zhang, 2019; Roy, 2020), which is lightweight. MQTT supports Quality of Service (QoS) levels to ensure message delivery and reliability, making it a good fit for low-power, low-bandwidth networks (Fongsamut, 2023).

HTTP is a common protocol for sending data over the internet. HTTP is used in DT manufacturing to access web-based interfaces and APIs (Qi, 2018; Fuller, 2020; Semeraro, 2021; Qi, 2021). Compared to the MQTT protocol, the HTTP protocol can transport huge amounts of data in small packets, which may result in higher bandwidth utilization. Data flow for the HTTP protocol is based on the architecture of client-server communication. HTTP protocol has the benefit of sending enormous amounts of data, but communication will be more delayed.

The Internet Engineering Task Force (IETF) created the Constrained Application Protocol (CoAP), a synchronous request/response application protocol, with a constrained resource device in mind. It was created utilizing a portion of the HTTP technique, allowing it to be compatible with HTTP (Karagiannis, 2015).

OPC UA is an open standard that offers a useful foundation for digitalization. It facilitates vertical communication from device to cloud and horizontal communication from machine to machine (Liu, 2023). Both CoAP and OPC UA protocols provide a standardized framework for communication between different devices, systems, and software applications (Lu, 2020; Redelinghuys, 2020). Point-to-point and client-server communication formats are supported by OPC UA, which makes it appropriate for a variety of applications.

6.4.4 IoT architecture

The IoT architecture refers to the structure and framework that enables the connection, communication, and interaction between various physical devices, sensors, actuators, and other objects embedded with software, sensors, and network connectivity (Soumyalatha, 2016; Al-Saedi, 2017; Alshohoumi, 2019).

The primary goal of IoT architecture is to facilitate the seamless integration of these devices into the internet and enable them to collect, exchange,

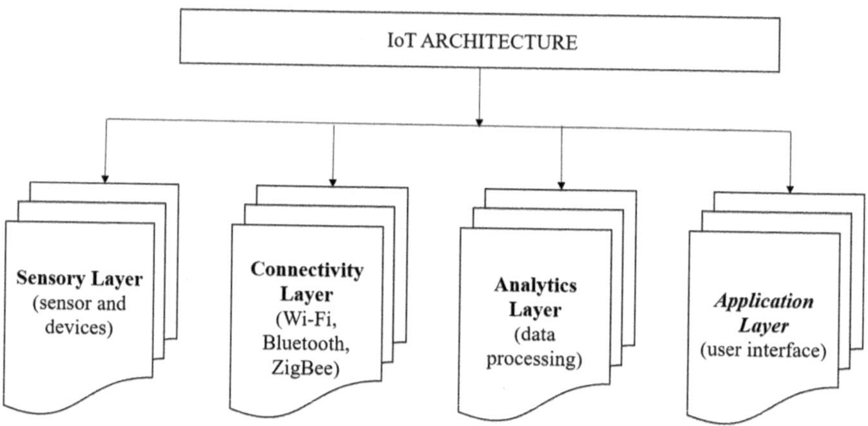

Figure 6.4 IoT architecture.

and analyze data in real time. This leads to more efficient processes, improved decision-making, and innovative services. IoT is a subpart of DT architecture; it plays a key role in developing a DT. Understanding its architecture improves clarity; thus, IoT architecture is also provided along with DT architecture. IoT architecture has four layers (Gokhale, 2018) as described in Figure 6.4.

Sensory or Perception layer: This layer includes the physical devices, sensors, and actuators that collect data from the environment or interact with it.

Connectivity layer: where the guidelines and conventions for data exchange between sensors, actuators, and edge devices in the sensing layer are established. This is achieved via wired or wireless communication protocols (Gokhale, 2018), or network modules such as Wi-Fi, Bluetooth, ZigBee, MQTT, CoAP, and Modbus are used in the IoT.

Analytics layer: where all the gathered unprocessed data from sensors, actuators and other sensing layer devices is transformed to digital format for processing and analyzing (Gokhale, 2018).

Application layer: where several application protocols like MQTT, CoAP, and AMQP that are created to satisfy the IoT requirements for low power consumption. These interfaces might be desktop programs, mobile applications, or web-based dashboards. Using charts, graphs, maps, and other graphical representations, users may monitor sensor readings, follow device status, and analyze analytics through the application layer's data visualization features (Ara, 2016). A knowledge of creating interfaces and dashboards is essential for making an interactive DT.

6.4.5 Dashboard

A dashboard is an aesthetically pleasing interface that displays data in real time, KPIs, and insight obtained from a DT (García Sánchez, 2021; Santra, 2021). In the context of tracking movement and gathering data from a physical twin, a dashboard is a graphical user interface that gives users an integrated view of both past and current data about the physical twin. The first step in building a dashboard is data integration. Designers and developers work together to create the dashboard structure and presentation tools when the data is gathered. Integrating a dashboard with its underlying data sources is the last stage of the creation process. Users can always rely on the dashboard to provide the most recent information. Various tools to create dashboards are MICROSOFT AZURE, R-LANGUAGE, MATLAB, SCALA, JAVA SCRIPT, AMAZON WEB SERVICE, etc.

6.5 CASE STUDY OF DEVELOPING DIGITAL TWIN FOR A MANUFACTURING SYSTEM

This section talks about the case study of developing a DT for a conventional CNC machine (Promill 8000 from Intellitek) to convert to a CPS. Detailed steps in development are given below.

6.5.1 Model development

To develop the DT of the CNC machine, CAD models of the machine's parts, such as tool bit, tool holder, and spindle, are created using AutoCAD software as shown in Figure 6.5.

The software that is used should be capable of animating the model in response to the real-time data from sensors. AutoCAD and other CAD software do not have the capability to generate motion with real-time data. So, the UNITY game engine software is used and has the capability to adopt real-time data and animate the physical part by providing appropriate information. It is a software development tool used for creating video games. CAD models of the CNC machine created in CAD software have to be imported to the UNITY 3D for creating DT. However, UNITY does not support the direct import of the machine's CAD models because of file extension compatibility problems (.ipt and .iam). These CAD models must be turned into .obj files, which UNITY can utilize as assets to be imported into the game.

Once the parts are imported into the UNITY, they become game objects (Zhu, 2019). As with real-world physical components of machines, these objects must be described with physics, such as gravity and movement limits. To accomplish this, algorithms are developed in the C# or C Sharp

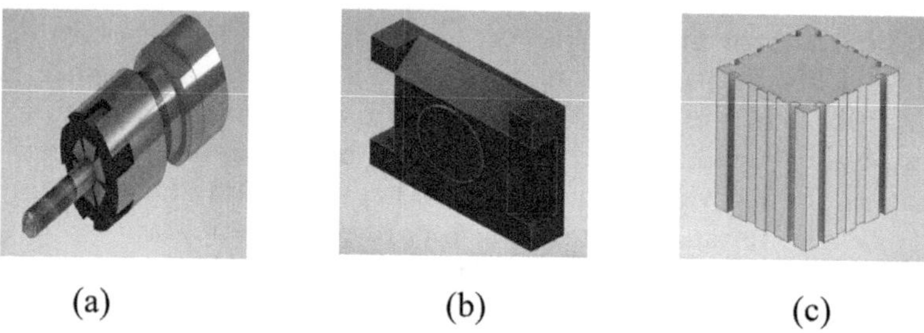

Figure 6.5 CAD models of CNC machine tool parts: (a) tool bit, (b) tool holder, and (c) spindle motor.

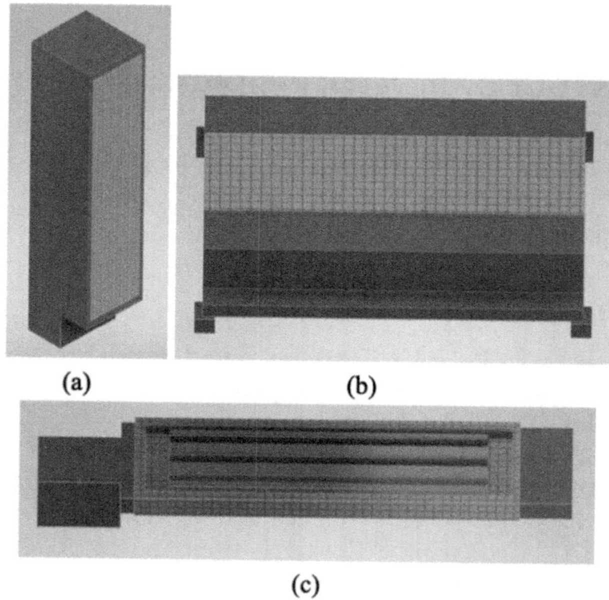

Figure 6.6 Converted .obj files of (a) machine column, (b) machine base, and (c) machine table (Borra, 2023).

language. C# scripts have the power to move the UNITY object. Unique interactions and activities in a gaming environment can be designed by using scripts. UNITY is used to run C# programs that are compiled in Microsoft Visual Studio.

Since visual representation was the main goal of the UNITY twin creation process, only the machine's exact measurements are considered. Only

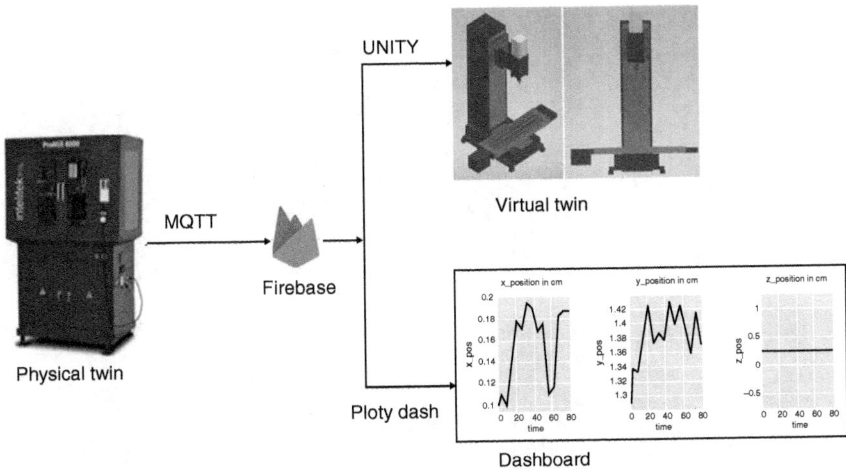

Figure 6.7 Architecture of digital twin created for CNC machine.

the .obj files of the externally visible components like the machine column, base, and table, as shown in Figure 6.6, are imported, allowing for a more efficient visualization process.

The machine column facilitates vertical movement along the Z-axis; the machine table can move along the X-axis and Y-axis over the machine base. The workpiece and clamp are usually fastened to the machine table, which enables horizontal movement along the X and Y axes. Every part imported has to be given rigid body characteristics, and constraints must be defined in UNITY based on the motions of CNC machine parts. Using the features of UNITY, the movement of the DT can be animated, allowing the game object of the CNC machine to travel in the direction of the desired position as shown in Figure 6.7.

6.5.2 Hardware integration

Hardware integration includes sensor integration on the CNC machine and interfacing sensors with the processing units. Important information, including coolant level, coolant temperature, spindle temperature, spindle current, door status, tool position, and cutting speed, cannot be retrieved by the present software in the Intelitek Promill 8000 machine. Various sensors, like temperature sensors (MLX90614), IR sensors, ultrasonic sensors, Hall effect sensors, and accelerometers, are installed on CNC machine, as shown in Figure 6.8.

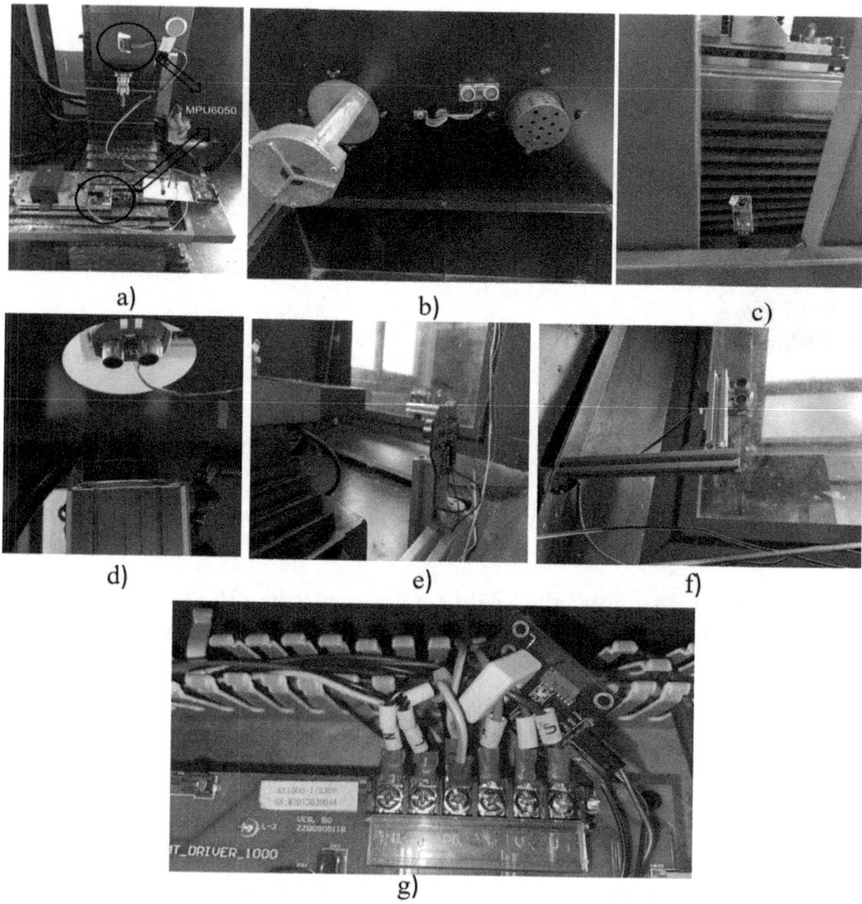

Figure 6.8 Integration of sensors: (a) sensor for cutting tool feed rate change, (b) sensor installation for coolant level and temperature, (c) IR sensor installation for monitoring the door status, (d) WCS1700 for detecting spindle power consumption, (e) ultrasonic sensor for measuring the vertical movement of the spindle, (f) ultrasonic sensor for measuring the longitudinal motion of the bed, and (g) ultrasonic sensor for measuring the lateral motion of the bed (Reddy, 2023).

6.5.3 Data acquisition

Once the sensors are integrated on the machine, there should be a processing unit, which can collect data. Raspberry Pi general purpose input/output (GPIO) pins are utilized for sensor interfacing. These pins can be set up to read digital or analog signals from the sensors and transform them into data that can be used for analysis.

6.5.4 Cloud interfacing

Firebase is an extensive platform of Google, which provides a wide range of services for building and managing mobile and internet-based programs. Data are sent to Firebase from Rpi using the Pyrebase library (Reddy, 2023). To send data from Firebase to UNITY, android SDK libraries are used. They may then access relevant data from specified database nodes and use UNITY-integrated graphics to create an immersive and dynamic data visualization.

6.5.5 Real-time monitoring

A dashboard is an easily observable and updatable visual depiction of the most important parameters of the system. Plotly is a premier open-source data visualization platform that offers a variety of dynamic and customizable chart types, including heatmaps, line charts, bar charts, and scatter plots. Developers may construct interactive internet apps with Plotly visualization (Vuppugalla, 2023) using the web-based framework Dash. The case study uses the Plotly library to create dashboards as it is open-source. Dashboards are created for updating tool position, door status, coolant temperature, and coolant level. Call-backs are made such that by pressing the button on the dashboard, the dashboard button feature allows the user to quickly contact operators and retrieve their contact information, as shown in Figure 6.9.

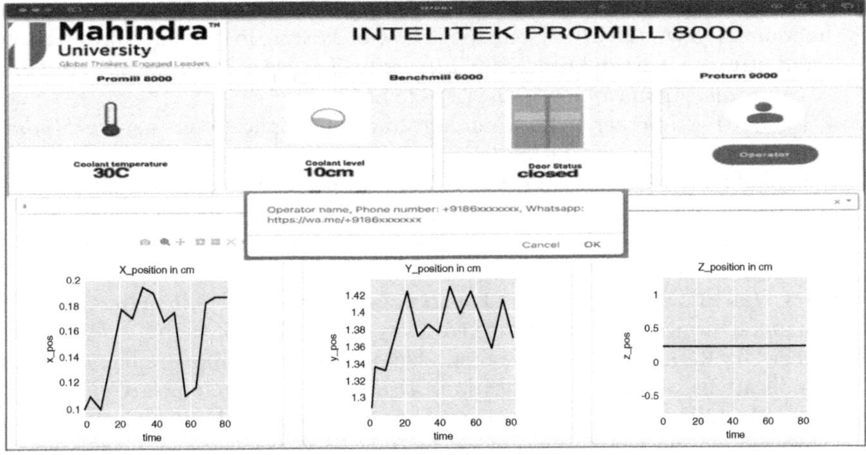

Figure 6.9 Real-time data monitoring through dashboards.

6.6 CONCLUSIONS AND FUTURE SCOPE

Research on DT has shown their enormous potential and effect in a variety of fields, especially in manufacturing. The chapter explained the definitions and applications of DT, demonstrating their adaptability and value in various industries. DT has the ability to improve manufacturing efficiency, human-machine interaction, fault diagnosis, and productivity due to its advanced features of data storage and integration with IoT and AI. DTs evolved from a level where they were implemented on a single object to the level of replicating system of systems, and they are able to influence the complete product life cycle. A complete architecture of DT is explained along with various tool to realize DT. A case study of developing DT for converting a CNC machine to CPS is discussed, explaining even the minute details.

Two-way communication between the digital and physical worlds holds immense potential for revolutionizing the manufacturing process. Along with improving productivity and operational efficiency, this feature creates new opportunities for remote decision-making, troubleshooting, and collaboration. This technology has the ability to increase the scope of DT.

REFERENCES

Agnusdei, G. P. (2021). Is digital twin technology supporting safety management? A bibliometric and systematic review. *Applied Sciences, 11(6)*, 2767.

Aliper, A. P. (2016). Deep learning applications for predicting pharmacological properties of drugs and drug repurposing using transcriptomic data. *Molecular Pharmaceutics, 13(7)*, 2524–2530.

Al-Saedi, I. R. (2017). CNC machine based on embedded wireless and Internet of Things for workshop development. In *2017 International conference on control, automation and diagnosis (ICCAD)* (pp. 439–444). IEEE.

Alshohoumi, F. S.-A. (2019). Systematic review of existing IoT architectures security and privacy issues and concerns. *International Journal of Advanced Computer Science and Applications, 10(7)*, 232–251.

Ara, T. S. (2016). Internet of Things architecture and applications: A survey. *Indian Journal of Science and Technology, 9, 1–7.*

Asavasirikulkij, C. M. (2022). A study of digital twin and its communication protocol in factory automation cell. In *Proceedings of the International Conference on Emerging Technologies for Communications* (pp. 1–3). Online ISSN:2188-5079, DOI:10.34385/proc.68.D4-2

Bao, J. G. (2018). The modelling and operations for the digital twin in the context of manufacturing. *Enterprise Information Systems, 13(4)*, 534–556.

Barricelli, B. R. (2019). A survey on digital twin: Definitions, characteristics, applications, and design implications. *IEEE Access, 7*, 167653–167671.

Borra, S. V. (2023). Developing digital twin for transforming CNC machine to Industry 4.0 for remote monitoring. In *2023 IEEE International Conference on Mechatronics and Automation (ICMA)* (pp. 100–105). IEEE.

Capra, R. V. (2014). File synchronization and sharing: User practices and challenges. *Proceedings of the American Society for Information Science and Technology, 51(1)*, 1–10.

Chen, Y. (2017). Integrated and intelligent manufacturing: Perspectives and enablers. *Engineering, 3(5)*, 588–595.

Chryssolouris, G. M. (2009). Digital manufacturing: history, perspectives, and outlook. *Proceedings of the Institution of Mechanical Engineers, Part B: Journal of Engineering Manufacture, 223(5)*, 451–462.

Cimino, C. N. (2019). Review of digital twin applications in manufacturing. *Computers in Industry, 113*, 103130.

Fang, L. L. (2021). A digital twin-oriented lightweight approach for 3d assemblies. *Machines, 9(10)*, 231.

Farsi, M. D.-F. (2020). *Digital Twin Technologies and Smart Cities (Vol. 1134)*. Springer International Publishing.

Feng, Y. T. (2021). Real-time and on-line monitoring of ethanol fermentation process by viable cell sensor and electronic nose . *Bioresources and Bioprocessing, 8(1)*, 37.

Fongsamut, C. K. (2023). Digital twin in automation industry: Optimal communication protocol. In *2023 7th International Conference on Information Technology (InCIT)* (pp. 33–37). IEEE.

Fuller, A. F. (2020). Digital twin: Enabling technologies, challenges and open research. *IEEE Access, 8*, 108952–108971.

García Sánchez, A. P. (2021). *Development of a dashboard for the evolution of COVID-19*. Thesis, Universidad complutense de madrid. https://hdl.handle.net/20.500.14352/10472

Ge, M. X. (2008). An intelligent online monitoring and diagnostic system for manufacturing automation. *IEEE Transactions on Automation Science and Engineering, 5(1)*, 127–139.

Ghorbani, Z. (2024). A categorical approach for defining digital twins in the AECO industry. *Journal of Information Technology in Construction, 29*, 198–218.

Glaessgen, E. (2012). The digital twin paradigm for future NASA and US Air Force vehicles. In *53rd AIAA/ASME/ASCE/AHS/ASC Structures, Structural Dynamics, Materials Conference, American Institute of Aeronautics and Astronautics* (p. 1818–1831).

Gokhale, P. B. (2018). Introduction to IOT. *International Advanced Research Journal in Science, Engineering and Technology, 5(1)*, 41–44.

Grieves, M. (2014). Digital twin: manufacturing excellence through virtual factory replication. *White Paper, 1(2014)*, 1–7.

Gunes, V. P. (2014). A survey on concepts, applications, and challenges in cyber-physical systems. *KSII Transactions on Internet and Information Systems (TIIS), 8(12)*, 4242–4268.

Harpe, P. M. (2021). *Analog Circuits for Machine Learning, Current/Voltage/Temperature Sensors, and High-speed Communication: Advances in Analog Circuit Design 2021*. Springer International Publishing.

Hribernik, K. A. (2006). The product avatar as a product-instance-centric information management concept. *International Journal of Product Lifecycle Management, 1(4)*, 367–379.

Hu, W. Z. (2021). Digital twin: A state-of-the-art review of its enabling technologies, applications and challenges. *Journal of Intelligent Manufacturing and Special Equipment, 2(1)*, 1–34.

Iwata, K. O. (1995). A modelling and simulation architecture for virtual manufacturing systems. *CIRP Annals, 44(1)*, 399–402.

Iwata, Y. M. (2017). Effect of sensor installation on the accurate measurement of soil water content. *European Journal of Soil Science, 68(6)*, 817–828.

Jayal, A. D. (2010). Sustainable manufacturing: Modeling and optimization challenges at the product, process and system levels. *CIRP Journal of Manufacturing Science and Technology, 2(3)*, 144–152.

Jevtić, A. (2023). *Digital twins: definition, application options in the product life cycle and marketing*. In *60th Ilmenau Scientific Colloquium*. Technische Universität Ilmenau.

Junnan, Z. J. (2022). A modeling method of complex assembly based on digital twin. *Procedia CIRP, 114*, 79–87.

Karagiannis, V. C.-G.-Z. (2015). A survey on application layer protocols for the internet of things. *Transaction on IoT and Cloud computing, 3(1)*, 11–17.

Kritzinger, W. K. (2018). Digital Twin in manufacturing: A categorical literature review and classification. *IFAC-PapersOnLine, 51(11)*, 1016–1022.

Li, X. W. (2021). Framework for manufacturing-tasks semantic modelling and manufacturing-resource recommendation for digital twin shop-floor. *Journal of Manufacturing Systems, 58*, 281–292.

Liu, H. X. (2022). Digital twin-driven machine condition monitoring: A literature review. *Journal of Sensors, 2022(1), 1–13*. https://doi.org/10.1155/2022/6129995.

Liu, X. J. (2023). A systematic review of digital twin about physical entities, virtual models, twin data, and applications. *Advanced Engineering Informatics, 55*, 101876.

Liu, Z. C. (2019). Data super-network fault prediction model and maintenance strategy for mechanical product based on digital twin. *IEEE Access, 7*, 177284–177296.

Lu, Y. L. (2020). Digital Twin-driven smart manufacturing: Connotation, reference model, applications and research issues. *Robotics and Computer-Integrated Manufacturing, 61*, 101837.

Madni, A. M. (2019). Leveraging digital twin technology in model-based systems engineering. *Systems, 7(1)*, 7.

Mawson, V. J. (2019). The development of modelling tools to improve energy efficiency in manufacturing processes and systems. *Journal of Manufacturing Systems, 51*, 95–105.

Meierhofer, J. W. (2020). The digital twin as a service enabler: From the service ecosystem to the simulation model. In *Exploring Service Science. 10th International Conference, IESS 2020* (pp. 347–359). Springer International Publishing.

Miotto, R. W. (2017). Deep learning for healthcare: Review, opportunities and challenges. *Briefings in Bioinformatics, 19(6)*, 1236–1246.

Moheimani, R. H. (2022). Recent advances on capacitive proximity sensors: From design and materials to creative applications. *Journal of Carbon Research, 8(2)*, 26.

Monostori, L. K. (2016). Cyber-physical systems in manufacturing. *CIRP Annals, 65(2)*, 621–641.

Myers, B. A. (1998). A brief history of human-computer interaction technology. *Interactions, 5(2)*, 44–54.

Negri, E. F. (2017). A review of the roles of digital twin in CPS-based production systems. *Procedia Manufacturing, 11*, 939–948.

Pang, J. Z. (2023). A verification-oriented and part-focused assembly monitoring system based on multi-layered digital twin. *Journal of Manufacturing Systems, 68*, 477–492.

Patel, S. P. (2012). A review of wearable sensors and systems with application in rehabilitation. *Journal of Neuroengineering and Rehabilitation, 9*, 1–17.

Plessis, A. (2016). Virtual world-physical world: What is the real world? *International Journal of Management Science and Business Administration, 2(6)*, 42–54.

Qi, Q. (2018). Digital twin and big data towards smart manufacturing and industry 4.0: 360 degree comparison. *IEEE Access, 6*, 3585–3593.

Qi, Q. T. (2021). Enabling technologies and tools for digital twin. *Journal of Manufacturing Systems, 58*, 3–21.

Qiao, Q. W. (2019). Digital twin for machining tool condition prediction. *Procedia CIRP, 81*, 1388–1393.

Reddy, V. A. (2023). Transformation of Industry 3.0 CNC machines to Industry 4.0 machines: Sensor selection and integration. In *2023 IEEE International Conference on Mechatronics and Automation (ICMA)* (pp. 285–290). IEEE.

Redelinghuys, A. J. (2020). A six-layer architecture for the digital twin: A manufacturing case study implementation. *Journal of Intelligent Manufacturing, 31(6)*, 1383–1402.

Ríos, J. H. (2015). Product avatar as digital counterpart of a physical individual product: Literature review and implications in an aircraft. *Transdisciplinary Lifecycle Analysis of Systems*, IOS Press, 657–666. doi:10.3233/978-1-61499-544-9-657

Rosen, R. V. (2015). About the importance of autonomy and digital twins for the future of manufacturing. *IFAC-PapersOnLine, 48(3)*, 567–572.

Roy, R. B. (2020). Digital twin: current scenario and a case study on a manufacturing process. *The International Journal of Advanced Manufacturing Technology, 107*, 3691–3714.

Santra, A. S. (2021). An extensible dashboard architecture for visualizing base and analyzed data. *arXiv preprint, arXiv:2106.05357.*

Semeraro, C. L. (2021). Digital twin paradigm: A systematic literature review. *Computers in Industry, 130*, 103469.

Shafto, M. C. (2010). Draft modeling, simulation, information technology & processing roadmap. *Technology Area, 11*, 1–32.

Shao, G. (2020). Framework for a digital twin in manufacturing: Scope and requirements. *Manufacturing Letters, 24*, 105–107.

Singh, M. F. (2021). Digital twin: Origin to future. *Applied System Innovation, 4(2)*, 36.

Soumyalatha, S. G. (2016). Study of IoT: understanding IoT architecture, applications, issues and challenges. In *1st International Conference on*

Innovations in Computing & Net-working (ICICN16), CSE, RRCE (Vol. 478). International Journal of Advanced Networking & Applications.

Sun, L. P. (2020). Dynamic analysis of digital twin system based on five-dimensional model. *Journal of Physics: Conference Series, 1486(7)*, 072038.

Tao, F. C. (2018). Digital twin-driven product design, manufacturing and service with big data. *The International Journal of Advanced Manufacturing Technology, 94*, 3563–3576.

Tao, F. Q. (2022). *Digital Twin Driven Service.* Academic Press.

Tao, F. Z. (2019). *Digital Twin Driven Smart Manufacturing.* Academic Press.

Tuegel, E. J. (2011). Reengineering aircraft structural life prediction using a digital twin. *International Journal of Aerospace Engineering, 2011(1), 1–14.* doi:10.1155/2011/154798

Uhlemann, T. H. (2017a). The digital twin: Demonstrating the potential of real time data acquisition in production systems. *Procedia Manufacturing, 9*, 113–120.

Uhlemann, T. H. (2017b). The digital twin: Realizing the cyber-physical production system for industry 4.0. *Procedia CIRP, 61*, 335–340.

Veverka, D. V. (2020). *The Application of Active and Passive Optical Sensors in Natural Resource Decision Making.* NDSU Libraries.

Vuppugalla, S. G. (2023). *Development of a smart interfacing dashboard between management and shop-floor for Industry 3.0 machines and small and medium enterprises (SME).* In *2023 IEEE International Conference on Mechatronics and Automation (ICMA)* (pp. 403–408). IEEE.

Wang, G. Z. (2021). Digital twin-driven service model and optimal allocation of manufacturing resources in shared manufacturing. *Journal of Manufacturing Systems, 59*, 165–179.

Warke, V. K. (2021). Sustainable development of smart manufacturing driven by the digital twin framework: A statistical analysis. *Sustainability, 13(18)*, 10139.

Wilhelm, J. P. (2021). Review of digital twin-based interaction in smart manufacturing: Enabling cyber-physical systems for human-machine interaction. *International journal of computer integrated manufacturing, 34(10)*, 1031–1048.

Wynn, M. (2023). Digital twin applications in manufacturing industry: A case study from a German multi-national. *Future Internet, 15(9)*, 282.

Yao, J. F. (2023). Systematic review of digital twin technology and applications. *Visual Computing for Industry, Biomedicine, and Art, 6(1)*, 10.

You, Y. C. (2022). Advances of digital twin for predictive maintenance. *Procedia Computer Science, 200*, 1471–1480.

Zhang, C. Z. (2019). A data-and knowledge-driven framework for digital twin manufacturing cell. *Procedia CIRP, 83*, 345–350.

Zheng, Y. Y. (2019). An application framework of digital twin and its case study. *Journal of Ambient Intelligence and Humanized Computing, 10*, 1141–1153.

Zhong, D. X. (2023). *Overview of Predictive Maintenance Based on Digital Twin Technology*. Heliyon.

Zhu, Z. L. (2019). Visualisation of the digital twin data in manufacturing by using augmented reality. *Procedia CIRP, 81*, 898–903.

Zou, Y. (2004). Sensor deployment and target localization in distributed sensor networks. *ACM Transactions on Embedded Computing Systems (TECS), 3(1)*, 61–91.

Chapter 7

Decision-making in Industry 4.0

Theophilus Dhyankumar C., Joseph Francis J., and Sivakumar K.

7.1 INTRODUCTION

Recent years have witnessed a surge in interest among manufacturing companies of all sizes towards Industry 4.0 and its associated technologies. The main objective of all these programs is to increase profitability and agility by connecting customers, machinery, products, and supply chains. These will ensure better decision-making capabilities across the entire system. Given the multitude of interpretations surrounding the concept of Industry 4.0, there remains ambiguity regarding the precise definition and delineation of its associated technologies (Dombrowski et al., 2017; Mayr et al., 2018). The Boston Consulting Group, for instance, arranged these technologies according to the nine Industry 4.0 fundamental pillars. Big data analysis, cybersecurity, cloud computing, horizontal and vertical integration of systems, Internet of Things (IoT), autonomous robotics, simulation, augmented reality (AR), and additive manufacturing are among the pillars. However, few authors have identified that there may be more than nine pillars (Sanders et al., 2016; Wagner et al., 2017; Moeuf et al., 2018). Thus, the domain of decision-making using Industry 4.0 is too vast, and hence classification concerning the application to different fields and the autonomy in decision-making is more useful, adhering to the number of available technologies.

There are two major differences between decision-making and Industry 4.0. Decision-making can be applied to implement Industry 4.0. Industry 4.0 technologies can be used to connect the above-mentioned nine pillars. Also, implementing Industry 4.0 requires substantial investment, a proficient workforce, and cutting-edge technologies (Müller et al., 2018). Silva et al. (2022) have pinpointed the primary criteria guiding decision-making processes in the selection and integration of technologies associated with the Industry 4.0 framework. Osterrieder et al. (2020) presented an investigational framework related to the smart factory. The research model provided eight thematic perspectives, one of which focused on decision-making. The model emphasized the pervasive nature of decision-making challenges, which

DOI: 10.1201/9781003470861-7

intersect with various thematic areas and impact a wide array of manufacturing activities. In fact, by streamlining problem-solving and other decision-making processes, Industry 4.0 technologies like artificial intelligence (AI), AR, cobotics, extensive data analysis, the IoT, and machine learning (ML) can improve the autonomy of production systems, including both operators and equipment. This is a key component of the idea of autonomous intelligent factories; thus, it is not unexpected that a lot of research has been done on data-driven production decision-making in design, planning, process control, and scheduling. However, the proposed models fail to effectively incorporate the many types of autonomy within the decision-making process made possible using all Industry 4.0-related technologies. The earliest decision-making model by Simon in the year 1960 provided a comprehensive depiction of a rational decision-making approach (Lin et al., 2012). Building upon this decision-making model, Mintzberg et al. (1976) offered a methodology for making strategic decisions in organizations. The research findings indicate a prevalence of broad requirements, encompassing diverse criteria aimed at aggregating various needs in choosing techniques and tools for decision-making in Industry 4.0. Thus, this chapter focuses on strategies that assist in making decisions from an Industry 4.0 viewpoint, as well as tools and techniques that might guide industrial practitioners and researchers in making Industry 4.0-related decisions. Especially, this chapter will give insights into decision-making in Industry 4.0 from various perspectives along with real-time case studies.

The remainder of the chapter is structured as in the following section. Section 7.2 presents the past studies in decision-making together with the approaches adopted. The decision-making in Industry 4.0 from various perspectives is discussed in Section 7.3. Effective decision-making in Industry 4.0 with case studies is discussed in Section 7.4. Sections 7.5 and 7.6 present the managerial implications and conclusion of the chapter.

7.2 LITERATURE REVIEW

This section discusses the previous research studies on decision-making in Industry 4.0. The databases included for the search are Google Scholar, EBSCO database, Emerald, Science Direct, Taylor & Francis, Scopus, and Web of Science. The chapter also included the conference proceedings for obtaining important insights into this growing field of research. The keywords used for the search were Industry 4.0, smart factory, smart manufacturing, decision-making, multi-criteria decision-making (MCDM), and human cyber-physical (HCP) systems. This chapter selected literature review as a methodology because most of the studies have utilized decision-making and assisted in decision-makers choosing and implementing Industry 4.0 technology. The general research methodology that has been adopted in this chapter is shown in Figure 7.1.

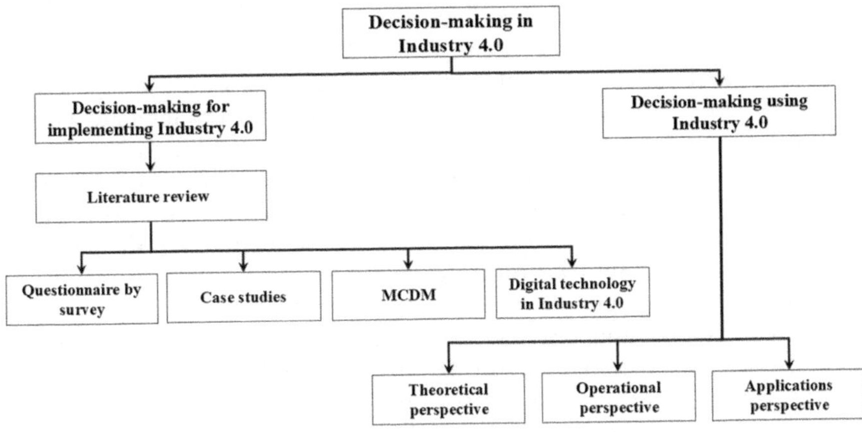

Figure 7.1 Flowchart for the research study (Source: Authors).

This section presents the past literature on how various tools and techniques were applied by researchers to make decisions in implementing Industry 4.0 with case studies, questionnaires by survey, and application of MCDM tools for decision-making and implementing digital technology in Industry 4.0.

For instance, Hamzeh et al. (2018) relied solely on a review of the literature to develop their conceptual framework, aimed at aiding decision-makers in the selection and adoption of Industry 4.0 technology. Similarly, Kaya et al. (2020) did a review of the literature to identify the factors used in technology selection and developed a schedule for assessing digital technology. Also, Amoozad Mahdiraji et al. (2020) employed a literature survey methodology in their research. Among the above-mentioned studies, the most popular technique for determining and categorizing the crucial and significant elements for Industry 4.0 deployment was the survey. This was demonstrated by Müller et al. (2018) and Türkeş et al. (2019) in their studies through understanding the specific variable used in the survey. They specifically used IBM SPSS Statistics software and Partial Least Squares-Path Modelling (PLS-SEM) to help with the analysis of the survey data. Further, a few authors used real-life situations using case studies (Erdogan et al., 2018; Sevinç et al., 2018; Erbay and Yıldırım, 2019; Beyaz and Yıldırım, 2020; Leone and Barni, 2020). Interestingly, Thomassen et al. (2014) employed the action research approach and developed it as a case study. Moreover, few authors in their studies have not mentioned the methodology adopted or the sources identified for real-life scenarios (Daneshjo et al., 2017; Hamzeh et al., 2018; Olfati et al., 2020).

Similarly, within the naturalistic decision-making (NDM) domain, efforts have been directed toward describing the actual decision-making process within operational contexts (Klein, 2008). Power et al. (2019) examined the biases and constraints inherent in human decision-making. Other decision-making models have surfaced in the field of AI and intelligent agent development that are modeled after human decision-making frameworks. Ansari et al. (2018) discussed how human-centered cyber-physical production systems helped in understanding how Industry 4.0 impacts decision-making tasks. Rosin et al. (2021) introduced a model to assess the influence of Industry 4.0 technologies on decision-making processes. The model of how various technologies are used for Industry 4.0 impacts the autonomy of teams for better and faster decision-making. The model was found to be more suitable for manufacturing sectors that had implemented lean manufacturing. It proposes a generic model encompassing various forms of autonomy that management can adopt in terms of decision-making. The choice of autonomous models depends on the stages of the decision-making process that they seek to strengthen by utilizing various Industry 4.0 technologies.

Multi-criteria decision-making techniques, particularly the analytical hierarchy process (AHP), are widely used in conjunction with other tools to form a hybrid methodology. Sevinç et al. (2018) used AHP along with an analytical network process (ANP) to identify the transition factors for implementing Industry 4.0. Erdogan et al. (2018) applied AHP and Fuzzy VIKOR to determine which decision-making approach is most effective when integrating digital technologies within a company. Demircan Keskin et al. (2019) employed AHP and TOPSIS to identify the requirements for the adoption of digital technologies. Quality Function Deployment (QFD) and Analytic Hierarchy Process (AHP) were employed by Erbay and Yildirim (2019) to rank the requirements for Industry 4.0 implementation in the automobile sector. AHP and Design Thinking were used by Leone and Barni (2020) to develop a roadmap for integrating digital technology. Olfati et al. (2020) combined linear programming problems with fuzzy logic, particularly incorporating triangular fuzzy numbers techniques to identify the best digital technology. Similarly, Beyaz and Yıldırım (2020) applied only the TOPSIS method to identify the best digital technology. Daneshjo et al. (2017) developed an algorithm for selecting digital devices for machinery and equipment using computational and heuristic approaches. Amoozad Mahdiraji et al. (2020) deployed the Best-Worst Method (BWM) and Time-Division Multiple-Access – Interactive Video Interface File (TODIM-IVIF) to find out and classify the most important criteria for Industry 4.0 implementation. All the studies mentioned above have defined their research framework and used the multicriteria decision-making methods at different stages or phases in their research for the implementation of Industry 4.0. However, in some studies, the decision-making process has been developed for implementing and transforming Industry 4.0 (Sajjad et al., 2022; Silva et al., 2022).

From the literature, it has been found that the recent studies on decision-making in Industry 4.0 are very minimal and also provide valuable insights into the criteria and methodologies employed in decision-making concerning Industry 4.0 technology adoption. Further, the extent to which new technologies can improve the decision-making process remains uncertain in existing literature. A study was conducted to identify and characterize the potential of Industry 4.0 technologies to improve the decision-making process. Findings reveal that cloud computing emerges as a fundamental element in augmenting decision-making across the board. Still, some technologies show great promise at particular stages of the decision-making process, such as IoT and simulation.

7.3 DECISION-MAKING USING INDUSTRY 4.0

In this section, how decision-making has been utilized in Industry 4.0 is explained from the perspectives of theoretical, operational, and application, and it also explores the intelligent quality control tools and decision support systems for the Industry 4.0 environment.

7.3.1 Decision-making in Industry 4.0 from the theoretical perspectives

Data and the methodology used for analyzing the data play a pivotal role in the decision-making process in the rapidly evolving landscape of the manufacturing environment. This environment is characterized by dynamic shifts in demand forecasts, the integration of multiple technologies, the need for complex capabilities, and the synchronization of end-to-end supply chains. Industry 4.0 elements influence both data availability, facilitated by technologies like the IoT and Digitization, and methodology, which is enhanced through smart data analytics and overarching cognitive technologies (Arsene and Constantin, 2019). In the current Industry 4.0 landscape, various shop floor management methods are being implemented, such as lean manufacturing, logistics optimization, IoT integration, smart manufacturing, cyber-physical systems, and AI (Tripathi et al. 2022).

7.3.1.1 Human-centered cyber-physical system in Industry 4.0 decision-making

The decision-making processes can be decentralized and are relevant to the ability of cyber systems to make simple decisions and be autonomous. Moreover, the way and the extent to which decision-making is carried out in industries involving both humans and the cyber-physical system appear to be challenging due to the ever-growing need for the technologies (Marr, 2016). The transferability of decision-making activities between humans and cyber-physical production systems (CPPS) appears to be a critical job, with an emphasis on understanding the complementing link between human and CPPS skills rather than considering them as substitutes. To achieve this, the researchers adopt hybrid approaches combining qualitative and quantitative models. In qualitative analysis, a vector of competence and autonomy (VCA) is established to gauge the collaboration between humans and CPPS in decision-making processes. This qualitative assessment helps to identify the extent to which humans and CPPS can effectively work together.

The quantitative study considers human elements such as operation/handling time, learnability rate, and error probability rate, as well as various configurations of digital assistance systems (DAS) with varying degrees of automation. The impact on decision-making tasks can be assessed by examining various combinations of technological components. Then the outcome of the quantitative analysis is utilized to instantiate the VCA. Through the application of predefined rules, VCA values are interpreted, the current level of complementarity between humans and CPPS is determined, and also the potential transitions are identified in terms of radical or incremental, which could lead to achieving the desired level of complementarity. This analysis is carried out from the TU Wien pilot factory Industry 4.0 context to let in real-world application and validation. Henceforth, the challenge in future decision-making processes lies in leveraging brain function as a predictive mechanism capable of utilizing mood, context, and social stimuli for active inference rather than attempting to suppress these capabilities (Barrett, 2017).

7.3.2 Decision-making in Industry 4.0 from the operational perspective

The field of decision-making encompasses research activities related to data-driven decision-making in manufacturing, utilizing various technologies such as visualization techniques, ML, and AI. This stream includes design, scheduling, process planning, control, and all other aspects of manufacturing decision-making. Various models are available to explain the different forms of autonomy and the role of Industry 4.0 technology in the different stages of the decision-making process. From a decision-making perspective, improvements in technologies favor implementing a new model of autonomy. The model can be used by decision-makers to comprehend the opportunities associated with the convergence of cybernetic, physical, and social spaces made feasible by Industry 4.0.

7.3.2.1 Machine learning-based prediction of overall equipment effectiveness

The most common way of approach to assessing process performance is based on key performance indicators (KPIs) (Zhang et al., 2018). However, it is important to acknowledge that this KPIs approach is used in ML techniques to examine the impact. Some of the applications of ML algorithms in overall equipment effectiveness (OEE) are discussed in brief. A hybrid analysis is carried out to locate bottleneck stations at the semi-automated car assembly line by combining human and clustering analysis (Dobra and Josvai, 2020). Similarly, Yu et al. (2019) studied neural networks to estimate and evaluate the efficiency loss caused by fifteen different individual input elements, such as process duration, usable tool, lot size standard deviation, etc. Brunelli et al. (2019) presented a deep learning method by analyzing production performance data about measures, alerts, and warnings for production performance predictions. Djatna and Alitu (2015) studied simulation to find a rule that shows the well-computed relationship between measurable indicators of OEE. Employing a case study focused on overall equipment effectiveness calculation within a manufacturing setting and subsequently analyzing the various traditional models and principles, most authors aim to assess the potential impact of Industry 4.0 elements on the decision-making process. Specifically, they seek to evaluate how Industry 4.0 elements may affect data collection and interpretation, alter the traditional approach to performance calculation, and integrate data into the calculation of performance indicators.

7.3.2.2 Optimization of process parameters using multi-criteria decision-making tools

Industrial engineering has also individually emerged as the leading field in terms of paper publications, surpassing other fields in the optimization of process parameters and applying multi-criteria decision-making methods. Multiple criteria decision-making (MCDM) techniques are used to make decisions when there are multiple criteria and no clear-cut answer. These techniques are commonly used in Industry 4.0 to make decisions related to supply chain management, production planning, and quality control. Some common applications of MCDM techniques in Industry 4.0 include selecting the best supplier, optimizing production processes, and determining the most efficient quality control measures. These techniques allow for more flexibility and adaptability in decision-making, which is essential in the fast-paced and ever-changing environment of Industry 4.0. Moreover, the literature review revealed that a large percentage of industrial engineering subfields were primarily concerned with problems commonly encountered in the supply chain. This finding is consistent with previous studies where they highlighted the prevalence of supply chain-related topics in TOPSIS application review papers (Behzadian et al., 2012). The decision-making methodologies have been found in various industrial applications such as robot selection (Bairagi et al., 2014; Chodha et al., 2022), facility location selection (Mokhtarian et al., 2014a, 2014b), selection of material in the sugar industry (Anojkumar et al., 2014, Anojkumar et al., 2015), logistics selection (Tadic et al., 2014; Behera and Beura, 2023), analyzing supplier processes (Hashemian et al., 2014; Ali et al., 2023), facility layout problem (Altuntas et al., 2014; Zha et al., 2020), and also in hospital services (Akdag et al., 2014; Alamoodi et al., 2023). Different modeling techniques have been presented in fuzzy choice-making applications and theories. Several appropriate ways have been provided for modeling decision assisting, and assistance is given in developing alternatives as they consider the complexity of the process. The individuals participating in the decision-making process, the intended goals, the information at hand, the amount of time available, and other factors all play a role in selecting a problem-solving strategy and a model.

7.3.2.3 Potential failure and defect diagnosis and analysis

The creation of fault prediction strategies through decision-making techniques has been the subject of several contributions in the literature (Lucantoni et al., 2022). Regression tree models have been proposed by some authors for failure classification (Sezer et al., 2018; Mohamed et al., 2019), while association rules techniques are used to identify hidden trends amid failures (Antomarioni et al., 2022). To predict breakdowns in complex systems, researchers have also looked into mathematical programming (Pisacane et al., 2021); nevertheless, deep learning prediction models for particular maintenance optimization have recently been presented with positive outcomes (Hesabi et al., 2022). Failure mode and effect analysis (FMEA) stands as a potent method for analyzing, identifying, and categorizing failures. Also, FMEA assesses the risks associated with these faults. Originating in the 1960s, this method found initial application in the aerospace industry to address quality and reliability issues in products. It was then embraced by the production sector as a risk assessment instrument to improve the stability and quality of the system. FMEA allows for the independent study of several phases of the product development process, including production. This allows for the identification and evaluation of potential fault types based on their risk and potential effects on subsequent manufacturing steps. When it comes to product manufacture, the kinds of errors that occur depend on the design, features, and capabilities of the production line. For this reason, professional assistance is required to determine machine dependencies and possible fault types. When it comes to product manufacture, the kinds of errors that occur depend on the design, features, and capabilities of the production line. For this reason, professional assistance is required to determine machine dependencies and possible fault types. Once the Risk Priority Numbers (RPNs) for each type of defect have been determined, the RPNs are assigned to the faults. These RPNs make it easier to compare the hazards connected to different types of machine malfunctions (Webert et al., 2022). In the above section, FMEA was introduced as a way to help professionals make decisions about machine hazards. Expert support techniques could be better researched and developed, for instance, through conducting in-depth expert interviews. Fault amendment is an undiscovered avenue. While some exciting progress has been made with automatically reconfiguring plants in a failure condition, there is still tremendous untapped potential in this subject. Future research in this area can be upon developing and validating non-statistical fault prioritizing techniques for Industry 4.0.

7.3.3 Decision-making in Industry 4.0 from the applications perspective

This section presents how data-driven decision-making in the future can assist Industry 4.0 maintenance applications. For instance, in the future, the decision-making may be arrived at by combining maintenance decision-making with other operational aspects like scheduling and planning, leveraging the cloud continuum for optimized deployment of decision-making services, improving methods for handling large-scale data, integrating advanced security measures, and connecting decision-making using simulation software, additive manufacturing, and autonomous robots. Moreover, AR and decision-making work together to seamlessly merge the physical and virtual realms of manufacturing operators.

7.3.4 Intelligent quality control tools for decision-making in Industry 4.0

In an Industry, and especially in a manufacturing or production system, quality is very important. Undoubtedly, the advent of novel techniques and production systems has necessitated the development of a fresh understanding of quality that emphasizes customized services and product design (Park, 1995). Many techniques for analyzing and monitoring quality management were used, starting with mass manufacturing, which was characterized by low diversity and vast volumes of items (Ngo and Schmitt, 2016). Quality management (QM) consequently gained popularity in the 1980s and 1990s, but businesses in the 21st century, during Industry 4.0, are still having difficulty implementing QM (Gunasekaran et al., 2019). Big data and Industry 4.0 for decision-making in quality control is another key area where a lot of research work has been conducted. As a result, four areas of focus have been identified that align with research needs: (1) identifying quality-pertinent data and sources; (2) designing an IT architecture; (3) using data mining techniques to analyze and predict quality improvements; and (4) developing quality control measures. These areas seek to establish a framework for quality management in Industry 4.0. Some of the specific applications of utilizing intelligent quality control in the Industry 4.0 scenario are presented in the subsequent section.

For instance, Irani et al. (2018) provide insight into managing organizational factors to minimize food waste using principles of design science. It delves into examining the causal relationships among consumption distribution factors. Kampker et al. (2018) employed Data-Use-Case-Matrix (DUCM) to analyze the data throughout the early stages of technological advancement in the automotive manufacturing sector. Kozjek et al. (2018) investigated manufacturing data using big data analytics to aid in operations planning through simulation that helps in anticipating potential resource overloads. The findings demonstrate that it can improve operational management, optimize resource use, and give decision-makers greater reliability. Para et al. (2019) presented the Analyze, Sense, Preprocess, Predict, Implement, and Deploy (ASPPID) technique in the automobile sector to find the faults in the annealing process with the aid of data analysts. Tsai et al. (2019) present a methodology for studying the relationship between the Activity-Based Standard Costing (ABSC) mixed decision model to generate optimal solutions and maximize profitability within resource restrictions. Studies on industry and manufacturing still make up the majority of the research. Nonetheless, research is now available looking for Industry 4.0 tools and methods to aid in decision-making in the fields of food, medicine, and sustainability (Goecks et al., 2020).

7.3.5 Decision support system for Industry 4.0 environment

For years, decision support systems and technologies have aimed to enhance human decision-making efficiency, bolster rational thinking, and mitigate prejudice, error, and bias. However, with Industry 4.0 advancements, neuroscience has made significant strides. Cognitive research now emphasizes implicit cognition, psychological and naturalistic processes, and the influence of social cues on human thought (Power et al., 2019). Consequently, decision support extends beyond rational models to encompass complex cognitive and analytical systems. Analysts or researchers involved in decision support must consider descriptive elements such as the traits, behaviors, and attitudes of users who interact with the analysis, systems, outputs, and outcomes (Kitchin, 2014; Ekbia et al., 2015). There has been considerable work exploring the conventional and modern perspectives on decision-making within the framework of adapting individual approaches to changing business environments, particularly in response to technological advancements associated with Industry 4.0 (Jankelova and Puhovichova, 2022). While the traditional emphasis on the rational aspect of decision-making remains prevalent, it is becoming increasingly overshadowed by the rapid advancements in computer technology and the availability of sophisticated software support. On the other hand, there's a growing trend toward inclusive decision-making procedures that incorporate all relevant parties, encourage group consensus, and value the capacity to develop and adjust in response to changing conditions and input. Interestingly, the evolving understanding of decision-making from its traditional conceptual boundaries has transcended to incorporating elements such as collective judgment and adaptability. This shift represents a fusion of rational decision-making with critical thinking and reasoning, resulting in a direction that is yet to be formally defined but embodies a synergy between traditional and modern approaches. The overall effectiveness of the decision-making process is influenced by various factors, including objective criteria established through a rational-normative model, environmental dynamics, and subjective influences such as individual personality traits and the cognitive complexity of the decision-maker.

The current findings indicate that integrating digital technologies into smart manufacturing alongside the proposed system has significantly enhanced production management efficiency and operational performance by leveraging smart systems, thereby promoting a safer shop floor management approach and improving financial standings. The decision-making system developed aims to optimize production sustainability while addressing constraints. Results from the investigation indicate that productivity has been enhanced through effective control of production activities on the shop floor. It presents a robust problem-solving framework aimed at revolutionizing Industry 4.0 methods, leading to increased productivity and benefiting industry stakeholders through improved smart shop floor management. Additionally, the study offers valuable perspectives and sustainable guidelines to assist industry professionals in implementing lean and smart manufacturing practices for productivity enhancement within the Industry 4.0 production environment (Tripathi et al., 2022). Another case study arose intending to consolidate a range of supply chain indicators into a system designed to enhance information management and transmission, thereby improving the decision-making process (Marques et al., 2020). Decision-making scientists, management experts, and information systems developers have yet to fully achieve the goal of reliably and efficiently constructing computer support systems for logical thinking, classical argumentation, or group decision-making. However, with the advent of advanced analytical tools, virtually limitless data repositories, and rapid distributed computing technology at our disposal, we can explore scientific advancements that are likely to unveil numerous aspects of human cognition that remain elusive (Power et al., 2019).

7.4 CASE STUDIES OF EFFECTIVE DECISION-MAKING IN INDUSTRY 4.0

7.4.1 Case study 1: The use of digital technologies among manufacturing firms

Numerous digital technologies that impact manufacturing firms in various situations are included in Industry 4.0 (Zheng et al., 2021). With the use of cutting-edge Industry 4.0 technologies, it is feasible to monitor the condition of manufacturing equipment, identify anomalies, and anticipate and resolve them before they arise (Jasiulewicz-Kaczmarek and Antosz, 2022). When properly examined, big data in this context frequently offers a wealth of information for manufacturing efficiency gains, cost reduction, and continuous improvement (Taghavi and Beauregard, 2020). Because it is simple to use and comprehend, MTBF (Main Time Between Failure) is the most widely used key performance indicator (KPI) in the asset assessment industry (Jittawiriyanukoon and Srisarkun, 2022). ML algorithms are frequently linked to asset performance, dependability, and MTBF, particularly when it comes to predicting maintenance schedule priority. The Deming cycle plus a ML framework make up the suggested methodology.

7.4.2 Case study 2: The use of machine learning framework and the Deming cycle in the overall equipment effectiveness

ML is used for the selection, implementation, and final assessment of the techniques to be used for overall equipment effectiveness. A three-step framework was proposed for implementing the rule-based ML methods that are used in the overall equipment effectiveness. The three processes are as follows: 1) data collection and processing for data management, overall equipment effectiveness, and collection; 2) identification and prioritization of major anomalies; and 3) giving the anomalies and hidden relationship investigations pertaining to OEE loss priority.

Plan, Do, Check, and Act is the Deming cycle designed to adopt proactive measures from an OEE continuous improvement standpoint. ML is used to detect anomalies. To find the relationships between the values and attributes kept in big datasets, association rule mining is utilized. Mining the association rules that link the urgent and significant failure modes with the appropriate OEE is the process of extracting knowledge. These case studies demonstrate the theoretical as well as the practical aspects of applying ML to OEE. Therefore, an inference can be made that ML techniques aid in the management of massive data sets and the decrease of anomalies in industrial processes, which in turn affects the efficiency of equipment.

7.4.3 Case study 3: The use of learning-based algorithm to combine various data mining techniques for fault prediction

Businesses employ data-driven strategies to help decision-makers manage vast amounts of data in a variety of contexts, including energy consumption (Mugnini et al., 2021), productivity in industrial operations (Antomarioni et al., 2021), efficiency (Görür et al., 2021), sustainability (Linke et al., 2019), and so forth. Both productivity and efficiency areas have shown the greatest potential for improvement in manufacturing due to the fusion of data analytics techniques and the advancement of information technology (Mansouri et al., 2020), primarily for failure detection in the field of maintenance (Görür et al., 2021). Before, the word "maintenance" was used to refer to time-based or breakdown-based maintenance plans, which gave fault occurrences a cyclical character while ignoring factors like stochastic unavailability because of a lack of reliable, high-quality data. To find hidden links between the occurrence of various failure modes (FMs), data mining techniques are applied to process observations, classify data, and perform analytics. For forecasting OEE labels and values, two distinct algorithms are utilized: Decision Tree J48 and Random Forest. Additionally, an Apriori algorithm is used to identify the optimal maintenance plan and detect interpretable patterns among the failure modes that have the biggest impact on OEE.

From the above cases, it is clear that manufacturing firms use digital technologies. Also, there are pieces of evidence of the use of a ML framework for improving the overall equipment effectiveness. Rule-based ML methods are used for the same. Also, learning algorithms among the various data mining techniques are used for fault prediction.

7.5 MANAGERIAL IMPLICATIONS OF DECISION-MAKING IN INDUSTRY 4.0

Academicians have shown a great deal of interest in Industry 4.0. The literature review from this study can serve as a guide for future investigation in decision-making for Industry 4.0. The results can be useful for top-level, middle-level, operational-level managers, and industrial policymakers. Research ought to be done in a variety of industries to determine how organizational strategy affects Industry 4.0 implementation success. Decision-making is key in any organization, and enhancing the implementation and effectively utilizing the data with the help of Industry 4.0 and its technologies remain crucial. Decision-making for implementing Industry 4.0 has to be classified and analyzed for a better understanding of the adaptability of the stakeholders. The technologies in Industry 4.0 also help in effective decision-making at the strategic, tactical, and operational levels in terms of application to products or processes. All the constituents and classifications are important in totality for successfully implementing and making the right decisions at the right time and right place in Industry 4.0.

7.6 CONCLUSION

The essential attribute that makes an organization triumphant is decision-making, and it also becomes a crucial role of every manager in the industry. The decision-making ability is a complicated phenomenon and depicts an interaction of reasonable and explanatory elements equally. According to the literature from past studies, there is a collective interest in getting an insight into decision-making and, in particular, a lot of researchers seek to analyze meticulously the concerned field of decision-making against the backdrop of Industry 4.0. Hence, in this chapter, extensive literature has been conducted to comprehend the importance of the decision-making process in Industry 4.0. Thus, the primary objective of this chapter is to direct the researchers and managers from the industry to find the tools and techniques that assist in decision-making in the Industry 4.0 scenario. Moreover, this chapter gave insights into how decision-making can be utilized from various perspectives together with intelligent quality control tools with the literature support. Further, the case studies discussed in this chapter will explore the evidence of the use of digital technologies among manufacturing firms, the use of ML frameworks in the era of the Deming cycle in overall equipment effectiveness, and the use of learning-based algorithms to combine various data-mining techniques for fault prediction. The literature analysis from this chapter can guide researchers to further analysis of decision-making in Industry 4.0 by selecting the proper tools and technologies that help in decision-making in real-time situations.

REFERENCES

Akdag, H., Kalaycı, T., Karagöz, S., Zülfikar, H., & Giz, D. (2014). The evaluation of hospital service quality by fuzzy MCDM. *Applied Soft Computing, 23*, 239–248.

Alamoodi, A. H., Albahri, O. S., Zaidan, A. A., Alsattar, H. A., Zaidan, B. B., & Albahri, A. S. (2023). Hospital selection framework for remote MCD patients based on fuzzy q-rung orthopair environment. *Neural Computing and Applications, 35*(8), 6185–6196.

Ali, M. R., Nipu, S. M. A., & Khan, S. A. (2023). A decision support system for classifying supplier selection criteria using machine learning and random forest approach. *Decision Analytics Journal, 7*, 100238.

Altuntas, S., Selim, H., & Dereli, T. (2014). A fuzzy DEMATEL-based solution approach for facility layout problem: a case study. *The International Journal of Advanced Manufacturing Technology, 73*, 749–771.

Amoozad Mahdiraji, H., Kazimieras Zavadskas, E., Skare, M., Rajabi Kafshgar, F. Z., & Arab, A. (2020). Evaluating strategies for implementing Industry 4.0: A hybrid expert oriented approach of BWM and interval valued intuitionistic fuzzy TODIM. *Economic Research-Ekonomska istraživanja*, *33*(1), 1600–1620.

Anojkumar, L., Ilangkumaran, M., & Sasirekha, V. (2014). Comparative analysis of MCDM methods for pipe material selection in sugar industry. *Expert Systems with Applications*, *41*(6), 2964–2980.

Anojkumar, L., Ilangkumaran, M., & Vignesh, M. (2015). A decision making methodology for material selection in sugar industry using hybrid MCDM techniques. *International Journal of Materials and Product Technology*, *51*(2), 102–126.

Ansari, F., Hold, P., & Sihn, W. (2018, June). Human-centered cyber physical production system: How does Industry 4.0 impact on decision-making tasks? In *2018 IEEE Technology and Engineering Management Conference (TEMSCON)* (pp. 1–6). IEEE.

Antomarioni, S., Ciarapica, F. E., & Bevilacqua, M. (2022). Association rules and social network analysis for supporting failure mode effects and criticality analysis: Framework development and insights from an onshore platform. *Safety Science, 150*, 105711.

Antomarioni, S., Lucantoni, L., Ciarapica, F. E., & Bevilacqua, M. (2021). Data-driven decision support system for managing item allocation in an ASRS: A framework development and a case study. *Expert Systems with Applications, 185*, 115622.

Arsene, C. G., & Constantin, G. (2019). Industry 4.0 and decision-making process. *Proceedings in Manufacturing Systems*, *14*(4), 129–134.

Bairagi, B., Dey, B., Sarkar, B., & Sanyal, S. (2014). Selection of robot for automated foundry operations using fuzzy multi-criteria decision making approaches. *International Journal of Management Science and Engineering Management*, *9*(3), 221–232.

Barrett, L. F. (2017). *How Emotions Are Made: The Secret Life of the Brain*. Pan Macmillan.

Behera, D. K., & Beura, S. (2023). Supplier selection for an industry using MCDM techniques. *Materials Today: Proceedings, 74*, 901–909.

Behzadian, M., Otaghsara, S. K., Yazdani, M., & Ignatius, J. (2012). A state-of the-art survey of TOPSIS applications. *Expert Systems with Applications*, *39*(17), 13051–13069.

Beyaz, H. F., & Yıldırım, N. (2020). A multi-criteria decision-making model for digital transformation in manufacturing: a case study from automotive supplier industry. In *Proceedings of the International Symposium for Production Research 2019* (pp. 217–232). Springer International Publishing.

Brunelli, L., Masiero, C., Tosato, D., Beghi, A. and Susto, G.A. (2019). Deep learning-based production forecasting in manufacturing: a packaging equipment case study. *Procedia Manufacturing*, *38*, 248–255, doi: 10.1016/j.promfg.2020.01.033.

Chodha, V., Dubey, R., Kumar, R., Singh, S., & Kaur, S. (2022). Selection of industrial arc welding robot with TOPSIS and entropy MCDM techniques. *Materials Today: Proceedings, 50*, 709–715.

Daneshjo, N., Majerník, M., Krivosudska, J., & Danishjoo, E. (2017). Modelling technical and economic parameters in selection of manufacturing devices. *TEM Journal*, *6*(4), 738.

Demircan Keskin, F., Kabasakal, İ., Kaymaz, Y., & Soyuer, H. (2019). An assessment model for organizational adoption of Industry 4.0 based on multi-criteria decision techniques. *Proceedings of the International Symposium for Production Research*, *18*, 85–100.

Djatna, T., & Alitu, I. M. (2015). An application of association rule mining in total productive maintenance strategy: An analysis and modelling in wooden door manufacturing industry. *Procedia Manufacturing*, *4*, 336–343, doi: 10.1016/j.promfg.2015.11.049.

Dobra, P., & Josvai, J. (2020). Enhance of OEE by hybrid analysis at the automotive semi-automatic assembly lines. *Procedia Manufacturing*, *54*, 184–190, doi: 10.1016/j.promfg.2021.07.028.

Dombrowski, U., Richter, T., & Krenkel, P. (2017). Interdependencies of Industrie 4.0 & Lean production systems: A use cases analysis. *Procedia Manufacturing*, *11*, 1061–1068.

Ekbia, H., Mattioli, M., Kouper, I., Arave, G., Ghazinejad, A., Bowman, T., Suri, V. R., Tsou, A., & Sugimoto, C. R. (2015). Big data, bigger dilemmas: A critical review. *Journal of the Association for Information Science and Technology*, *66*(8), 1523–1545.

Erbay, H., & Yıldırım, N. (2019). Technology selection for digital transformation: A mixed decision making model of AHP and QFD. *Proceedings of the International Symposium for Production Research*, *18*, 480–493).

Erdogan, M., Ozkan, B., Karasan, A., & Kaya, I. (2018). Selecting the best strategy for Industry 4.0 applications with a case study. In *Industrial Engineering in the Industry 4.0 Era: Selected Papers from the Global Joint Conference on Industrial Engineering and Its Application Areas, GJCIE 2017, July 20–21, Vienna, Austria* (pp. 109–119). Springer International Publishing.

Goecks, L. S., Santos, A. A. D., & Korzenowski, A. L. (2020). Decision-making trends in quality management: A literature review about Industry 4.0. *Production*, *30*, e20190086.

Görür, O. C., Yu, X., & Sivrikaya, F. (2021). Integrating predictive maintenance in adaptive process scheduling for a safe and efficient industrial process. *Applied Sciences*, *11*(11), 5042.

Gunasekaran, A., Subramanian, N., & Ngai, W. T. E. (2019). Quality management in the 21st-century enterprises: Research pathway towards Industry 4.0. *International Journal of Production Economics*, *207*, 125–129. http://dx.doi.org/10.1016/j.ijpe.2018.09.005.

Hamzeh, R., Zhong, R., Xu, X. W., Kajáti, E., & Zolotova, I. (2018, August). A technology selection framework for manufacturing companies in the context of Industry 4.0. In *2018 World Symposium on Digital Intelligence for Systems and Machines (DISA)* (pp. 267–276). IEEE.

Hashemian, S. M., Behzadian, M., Samizadeh, R., & Ignatius, J. (2014). A fuzzy hybrid group decision support system approach for the supplier evaluation process. *The International Journal of Advanced Manufacturing Technology*, *73*, 1105–1117.

Hesabi, H., Nourelfath, M., & Hajji, A. (2022). A deep learning predictive model for selective maintenance optimization. *Reliability Engineering & System Safety, 219*, 108191.

Irani, Z., Sharif, A. M., Lee, H., Aktas, E., Topaloğlu, Z., van't Wout, T., & Huda, S. (2018). Managing food security through food waste and loss: Small data to big data. *Computers & Operations Research, 98*, 367–383.

Jankelova, N., & Puhovichova, D. (2022). Managerial decision-making in the era of Industry 4.0. *Industry 4.0*, 7(2), 71–75.

Jasiulewicz-Kaczmarek, M., & Antosz, K. (2022, June). Industry 4.0 technologies for maintenance management – An overview. In *International Conference Innovation in Engineering* (pp. 68–79). Springer International Publishing.

Jittawiriyanukoon, C., & Srisarkun, V. (2022). Simulation for predictive maintenance using weighted training algorithms in machine learning. *International Journal of Electrical & Computer Engineering*, 12(3), 2839–2846.

Kampker, A., Heimes, H., Bührer, U., Lienemann, C., & Krotil, S. (2018). Enabling data analytics in large scale manufacturing. *Procedia Manufacturing, 24*, 120–127.

Kaya, İ., Erdoğan, M., Karaşan, A., & Özkan, B. (2020). Creating a road map for Industry 4.0 by using an integrated fuzzy multicriteria decision-making methodology. *Soft Computing*, *24*(23), 17931–17956.

Kitchin, R. (2014). Big data, new epistemologies and paradigm shifts. *Big Data & Society*, *1*(1), 1–12.

Klein, G. (2008). Naturalistic decision making. *Human Factors*, *50*(3), 456–460.

Kozjek, D., Rihtaršič, B., & Butala, P. (2018). Big data analytics for operations management in engineer-to-order manufacturing. *Procedia CIRP, 72*, 209–214.

Leone, D., & Barni, A. (2020, August). Industry 4.0 on demand: A value driven methodology to implement Industry 4.0. In *IFIP International Conference on Advances in Production Management Systems* (pp. 99–106). Springer International Publishing.

Lin, H. W., Nagalingam, S. V., Kuik, S. S., & Murata, T. (2012). Design of a global decision support system for a manufacturing SME: Towards participating in collaborative manufacturing. *International Journal of Production Economics*, *136*(1), 1–12.

Linke, B. S., Garcia, D. R., Kamath, A., & Garretson, I. C. (2019). Data-driven sustainability in manufacturing: Selected examples. *Procedia Manufacturing, 33*, 602–609.

Lucantoni, L., Ciarapica, F. E., &Bevılacqua, M. (2022). Learning-based algorithm for fault prediction combining different data mining techniques: A real case study. *The Eurasia Proceedings of Science Technology Engineering and Mathematics, 21*, 55–63.

Mansouri, S., Castronovo, F., & Akhavian, R. (2020). Analysis of the synergistic effect of data analytics and technology trends in the AEC/FM industry. *Journal of Construction Engineering and Management*, *146*(3), 04019113.

Marques, R., Moura, A., & Teixeira, L. (2020, August). Decision support system for the Industry 4.0 environment: Design and development of a business intelligence tool. In *Proceedings of the International Conference on Industrial Engineering and Operations Management.* Detroit, Michigan, USA, 1613–1624.

Marr, B. (2016). *What everyone must know about Industry 4.0. Forbes*. www.forbes.com/sites/bernardmarr/2016/06/20/what-everyone-must-know-about-industry-4-0/#:~:text=For%20a%20factory%20or%20system,in%20order%20to%20contextualize%20information

Mayr, A., Weigelt, M., Kühl, A., Grimm, S., Erll, A., Potzel, M., & Franke, J. (2018). Lean 4.0 – A conceptual conjunction of lean management and Industry 4.0. *Procedia CIRP, 72*, 622–628.

Mintzberg, H., Raisinghani, D., & Theoret, A. (1976). The structure of "unstructured" decision processes. *Administrative Science Quarterly*, *21*(2), 246–275.

Moeuf, A., Pellerin, R., Lamouri, S., Tamayo-Giraldo, S., & Barbaray, R. (2018). The industrial management of SMEs in the era of Industry 4.0. *International Journal of Production Research*, *56*(3), 1118–1136.

Mohamed, A., Hamdi, M. S., & Tahar, S. (2019). Decision tree-based approach for defect detection and classification in oil and gas pipelines. In *Proceedings of the Future Technologies Conference (FTC)*, *1*, 490–504.

Mokhtarian, M. N., Sadi-Nezhad, S., & Makui, A. (2014a). A new flexible and reliable IVF-TOPSIS method based on uncertainty risk reduction in decision making process. *Applied Soft Computing, 23*, 509–520.

Mokhtarian, M. N., Sadi-Nezhad, S., & Makui, A. (2014b). A new flexible and reliable interval valued fuzzy VIKOR method based on uncertainty risk reduction in decision making process: An application for determining a suitable location for digging some pits for municipal wet waste landfill. *Computers & Industrial Engineering, 78*, 213–233.

Mugnini, A., Polonara, F., & Arteconi, A. (2021). Energy flexibility curves to characterize the residential space cooling sector: The role of cooling technology and emission system. *Energy and Buildings, 253*, 111335.

Müller, J. M., Kiel, D., & Voigt, K. I. (2018). What drives the implementation of Industry 4.0? The role of opportunities and challenges in the context of sustainability. *Sustainability*, *10*(1), 247.

Ngo, Q. H., & Schmitt, R. H. (2016). A data-based approach for quality regulation. *Procedia CIRP*, *57*, 498–503. http://dx.doi.org/10.1016/j. procir.2016.11.086.

Olfati, M., Yuan, W., & Nasseri, S. H. (2020). An integrated model of fuzzy multi-criteria decision making and stochastic programming for the evaluating and ranking of advanced manufacturing technologies. *Iranian Journal of Fuzzy Systems, 17*(5), 183–196.

Osterrieder, P., Budde, L., & Friedli, T. (2020). The smart factory as a key construct of Industry 4.0: A systematic literature review. *International Journal of Production Economics, 221*, 107476.

Para, J., Del Ser, J., Nebro, A. J., Zurutuza, U., & Herrera, F. (2019). Analyze, sense, preprocess, predict, implement, and deploy (ASPPID): An incremental methodology based on data analytics for cost-efficiently monitoring the Industry 4.0. *Engineering Applications of Artificial Intelligence, 82*, 30–43.

Park, S. H. (1995). A new method of analysis for parameter design in quality engineering. *Total Quality Management*, *6*(1), 13–20.

Pisacane, O., Potena, D., Antomarioni, S., Bevilacqua, M., Emanuele Ciarapica, F., & Diamantini, C. (2021). Data-driven predictive maintenance policy based on multi-objective optimization approaches for the component repairing problem. *Engineering optimization*, *53*(10), 1752–1771.

Power, D. J., Cyphert, D., & Roth, R. M. (2019). Analytics, bias, and evidence: The quest for rational decision making. *Journal of Decision Systems*, *28*(2), 120–137.

Rosin, F., Forget, P., Lamouri, S., & Pellerin, R. (2021). Impact of Industry 4.0 on decision-making in an operational context. *Advances in Production Engineering & Management*, *16*(4), 500–514.

Sajjad, A., Ahmad, W., & Hussain, S. (2022). Decision-making process development for Industry 4.0 transformation. *Advances in Science and Technology. Research Journal*, 16(3), 1–11.

Sanders, A., Elangeswaran, C., & Wulfsberg, J. P. (2016). Industry 4.0 implies lean manufacturing: Research activities in Industry 4.0 function as enablers for lean manufacturing. *Journal of Industrial Engineering and Management (JIEM)*, *9*(3), 811–833.

Sevinç, A., Gür, Ş., & Eren, T. (2018). Analysis of the difficulties of SMEs in Industry 4.0 applications by analytical hierarchy process and analytical network process. *Processes*, *6*(12), 264.

Sezer, E., Romero, D., Guedea, F., Macchi, M., & Emmanouilidis, C. (2018, June). An Industry 4.0-enabled low cost predictive maintenance approach for SMEs. In *2018 IEEE International Conference on Engineering, Technology and Innovation (ICE/ITMC)* (pp. 1–8). IEEE.

Silva, J. F. D., Silva, F. L. D., Silva, D. O. D., Rocha, L. A. O., & Ritter, Á. M. (2022). Decision making in the process of choosing and deploying Industry 4.0 technologies. *Gestão & Produção, 29*, e163.

Tadić, S., Zečević, S., & Krstić, M. (2014). A novel hybrid MCDM model based on fuzzy DEMATEL, fuzzy ANP and fuzzy VIKOR for city logistics concept selection. *Expert Systems with Applications*, *41*(18), 8112–8128.

Taghavi, V., & Beauregard, Y. (2020, August). The relationship between lean and Industry 4.0: Literature review. In *Proceedings of the 5th North American Conference on Industrial Engineering and Operations Management*, Detroit, MI, USA (pp. 10–14).

Thomassen, M. K., Sjøbakk, B., & Alfnes, E. (2014). A strategic approach for automation technology initiatives selection. In *Advances in Production Management Systems. Innovative and Knowledge-Based Production Management in a Global-Local World: IFIP WG 5.7 International Conference, APMS 2014, Ajaccio, France, September 20–24, 2014, Proceedings, Part III* (pp. 288–295). Springer Berlin Heidelberg.

Tripathi, V., Chattopadhyaya, S., Mukhopadhyay, A. K., Saraswat, S., Sharma, S., Li, C., & Rajkumar, S. (2022). Development of a data-driven decision-making system using lean and smart manufacturing concept in Industry 4.0: A case study. *Mathematical Problems in Engineering*, *2022*, 1–20.

Tsai, W. H., Lan, S. H., & Huang, C. T. (2019). Activity-based standard costing product-mix decision in the future digital era: Green recycling steel-scrap material for steel industry. *Sustainability*, *11*(3), 899.

Türkeş, M. C., Oncioiu, I., Aslam, H. D., Marin-Pantelescu, A., Topor, D. I., & Căpuşneanu, S. (2019). Drivers and barriers in using Industry 4.0: A perspective of SMEs in Romania. *Processes*, *7*(3), 153.

Wagner, T., Herrmann, C., & Thiede, S. (2017). Industry 4.0 impacts on lean production systems. *Procedia CIRP, 63*, 125–131.

Webert, H., Döß, T., Kaupp, L., & Simons, S. (2022). Fault handling in Industry 4.0: Definition, process and applications. *Sensors*, *22*(6), 2205.

Yu, C.-M., Kuo, C.-J., Chiu, C.-L., Wen, W.-C., & Zhang, M. (2019). Unveil the black box for performance efficiency of OEE for semiconductor wafer fabrication. *IEEE International Symposium on Semiconductor Manufacturing Conference Proceedings*, 1–4. doi: 10.1109/ISSM.2018. 8651146.

Zha, S., Guo, Y., Huang, S., & Wang, S. (2020). A hybrid MCDM method using combination weight for the selection of facility layout in the manufacturing system: A case study. *Mathematical Problems in Engineering, 2020*, 1–16.

Zhang, K., Shardt, Y. A. W., Chen, Z., Yang, X., Ding, S. X., & Peng, K. (2018). A KPI-based process monitoring and fault detection framework for large-scale processes. *ISA Transactions*, *68*, 276–286, doi: 10.1016/j.isatra.2017.01.029.

Zheng, T., Ardolino, M., Bacchetti, A., & Perona, M. (2021). The applications of Industry 4.0 technologies in manufacturing context: A systematic literature review. *International Journal of Production Research*, *59*(6), 1922–1954.

Chapter 8

Navigating the legal landscape of sustainable Industry 4.0

Challenges and considerations

Soumya Prakash Patra

8.1 INTRODUCTION

As often, Industry 4.0, also called the fourth industrial revolution, understands it and is, in fact, the massive shift in industrial development and integration of advanced technologies in manufacturing and other industrial operations, a complete paradigm shift involving the convergence of cyber-physical systems, the IoT, AI, robotics, and big data analytics (Ghobakhloo, 2020). These innovations enable smart factories and connected systems that allow data collection, analysis, and decision-making to occur in real time. The key objective of Industry 4.0 is to increase productivity, reduce costs, and implement more responsive and adaptive manufacturing processes (Guo et al., 2021). This revolution increases the efficiency of mass production and opens a new chapter in industrial innovation and competition.

Hard on the heels of the technological breakthroughs of Industry 4.0, the urgency of sustainability is higher in the modern age of industry (Penna & Geels, 2012). Here is how each pillar devolves into the approach: The systematic approach enshrines sustainability – i.e., the proliferation of practices and technologies that are environmentally sound, economically feasible, and socially acceptable (Lasi et al., 2014). The urgency to address climate change, resource depletion, and social equity has led to incorporating sustainable development principles into industry operational strategies. Economic sustainability must go hand in hand with regulatory compliance, long-term business longevity, and brand reputation. Some of these practices include reducing waste, optimizing energy and resource use, minimizing greenhouse gas emissions, and promoting fair labor practices (Dauvergne & Lister, 2013). It poses crucial legal challenges to integrating sustainable practices into Industry 4.0.

Companies involved with Industry 4.0 frequently experience different challenges related to data privacy and security, intellectual property rights, and environmental compliance (Paez & Tobitsch, 2017). With integrated sensors and monitors, new technology's digital engagement can result in an unprecedented degree of connected devices and data analytics, and

DOI: 10.1201/9781003470861-8

these kinds of data privacy and protection remain a significant problem. Constraints like the General Data Protection Regulations adopted by the European Union impose stringent restrictions on data collection, storage, and processing, meaning they must have robust data protection solutions. In addition, it is challenging to adapt the existing intellectual property law because of the collaborative nature of Industry 4.0, meaning that it needs to expand the existing IP regime to safeguard the investments made in innovation through collaboration (Soares & Kauffman, 2018). Complex Intellectual Property circuits add an entirely new facet to these challenges, especially as multiple stakeholders are often involved in creating and implementing new technologies.

Another complexity stems from environmental regulations that emphasize the need to leave an ecologically friendly footprint, so a sustainable Industry 4.0 legal landscape also contributes to this area. To comply with international agreements like the Paris Agreement and national and regional environmental laws, industries need to embrace cleaner technologies and minimize their environmental impact (Hadi, 2024). Imposing regulations on industrial emissions (the one that comes to mind is the EU's Emissions Trading System – the ETS) forces industries to not only tweak their innovations (intuitive products for compliance) but to strive to permanently comply with higher environmental standards. Ad hoc changes in EU regulations lead to a mix of rules and approaches, making it challenging to anticipate the future of transportation and promote the correct use of emerging technologies that meet legal and sustainable demands.

This article examines these and other complex legal issues. It offers some ideas on strategic legal analysis and potential solutions for industries aiming to integrate sustainability and economics in the context of Industry 4.0. The preamble to my research article is intended and formalized into several headings with pure understanding: definitions, components, and principles of Industry 4.0 and sustainability, which have been detailed comprehensively in the background and context section. This will be followed by analyzing key legal issues, including data privacy, intellectual property, and environmental regulation. The article will then consider, practically, legal compliance and risk management. Real-world case studies will present examples of players successfully navigating the complex legal landscape. This article will also consider what the future may hold, exploring long-term consequences for industries and the legal landscape, as well as future trends and regulatory chronicles (Pinheiro et al., 2019).

8.2 BACKGROUND AND CONTEXT

8.2.1 What is Industry 4.0? Key components

The fourth industrial revolution, or Industry 4.0, is a game changer integrating new technologies into the manufacturing and industrial sectors. At the heart of Industry 4.0 is digitalizing and automating longstanding manufacturing processes, creating "smart factories" where machines and systems communicate and work together independently (Sony & Naik, 2019).

Industry 4.0 key elements:

1. Cyber-Physical Systems (CPS) are the integrations of computation, networking, and physical processes. Computational parameter synthesis consists of the computation of computational algorithms that allow real-time monitoring and control of a physical process (Jazdi, 2014).
2. Internet of Things (IoT): IoT links all devices and systems to collect and share data. With Industry 4.0, the IoT allows sensors and devices to collect data from industrial processes to improve productivity and predictive maintenance (Khan & Javaid, 2022).
3. Artificial Intelligence (AI) and Machine Learning (ML): AI and ML algorithms analyze the data for operations and predict failure and the best possible decision process. Such technologies drive these smart automation and adaptive production systems (Rai et al., 2021).
4. Big Data and Analytics: Processing and analyzing large amounts of data provides insights into operations performance, supply chain efficiency, and customer preferences, making decision-making more informed (Gokalp et al., 2016).
5. Cloud Computing: A cloud-based platform allows scalable storage and computation capabilities to process data in real time and collaborate with different parts of the world (Azadi et al., 2023).
6. Robotics and Autonomous Systems: The next generation of robotics automates mundane tasks and works with workers to increase productivity and minimize errors. Sustainability Principles for Industry 4.0 (Kovács et al., 2018).

To achieve this in the context of Industry 4.0, we need to integrate environmental, economic, and social aspects of sustainability into industrial practices. Main sustainability themes:

1. Resource Efficiency: Reducing waste and promoting recycling to conserve resources, especially energy, water, and materials. Circular economy principles are where the re-utilization and recycling of resources are internalized (Ghobakhloo, 2020).
2. Emission Reduction: Use of technologies and processes to reduce the production of greenhouse gases. This is attainable through energy-efficient machinery, utilization of renewable energy sources, and various carbon capture technologies (Ghobakhloo, 2020).
3. Life Cycle Assessment (LCA): An analysis to assess the environmental impacts associated with all stages of the life cycle of a product/process from the cradle to the grave (i.e., from raw material extraction

through materials processing, manufacture, distribution, use, repair and maintenance, and disposal or recycling) (Ferrari et al., 2021).

4. Social Responsibility – Fair labor, safe working environment, and community welfare. This includes ethical supply chain management and corporate social responsibility programs (Ocieczek & Gajdzik, 2019).
5. Financial Sustainability: The transfer of financial resources across generations while minimizing transfers related to the built, social and natural environments that balance the needs of communities, the economy, and the ecosystem to maintain their long-term health and integrity. This includes investment in the solutions that provide the returns, such as the benefits of sustainable technology (Soni et al., 2022).

8.3 LEGAL CHALLENGES IN SUSTAINABLE INDUSTRY 4.0

GDPR is one of the most critical data protection laws internationally, which can be viewed as a "golden standard." The General Data Protection Regulation is a regulation enforced by the European Union meant to protect the personal data of EU citizens. However, regardless of country, it applies to any organization processing that data. The main provisions include that people must give explicit consent for their data to be collected, that individuals currently have the right to review, update, and delete data they gave previously, and that severity levels be ordered to notify about a data breach (Rustad & Koenig, 2019).

Why GDPR is mandatory for companies using Industry 4.0 technologies, which require collecting and processing a lot of data. Failure to comply with this regulation may have serious consequences and can lead to fines of up to 4% of the total global turnover or €20 million, whichever is greater. Like GDPR, other authorities have put regulations in place, such as the California Consumer Privacy Act (CCPA) within the United States, that give consumers rights over their personal information and place duties on businesses to ensure data privacy (Schwartz, 2019).

8.3.1 Effects on IoT/big data analytics

The convergence of IoT devices and big data analytics in Industry 4.0 leads to serious problems regarding data security and privacy. The reality is that IoT devices collect enormous amounts of data from various sources, including machinery, sensors, and even human operators. This data are highly wanted for real-time monitoring, predictive maintenance, and optimization of industrial processes. The data that are collected are massive in scale and are often very personal, which correlate with the above. However, while this approach has served the objective of bettering many work processes and has improved many outputs, it has caused some very serious privacy and security issues. This function, however, serves the development team as they have to make sure the data they use is safe. That is helped by the use of strong encryption and making sure data get properly secured and access to the data controlled. The other is utilizing organized data management every step of the way, from data collection to data deletion. In order to comply with GDPR and similar legislation, industries have to conduct data protection impact assessments to find and address privacy risks that operations concerning IoT and big data analytics may cause (Khan et al., 2017).

8.3.2 Intellectual property (IP) copyright issues

Intellectual Property Rights in Collaborative Environments Industry 4.0 stimulates cooperation between manufacturers, technology providers, and investigative institutions. Working across these heterogeneous environments is imperative for cultivating ground-breaking solutions; notwithstanding, it additionally brings intricacy regarding IP. Proprietorship and rights over jointly engineered innovations can be contentious issues, with the potential result of disagreements suffocating chances for progress. Organizations must address these tests by establishing IP arrangements amid each brief before teaming up with others. These arrangements should lay out every party's respective entitlements and obligations concerning the exploration, advancement, commercialization, and utilization of the consequent innovations (Rahanu et al., 2021).

Moreover, companies must investigate open innovation platforms that permit IP sharing while shielding all stakeholders' interests. Patent Law Safeguards for Clean Technologies In high-tech manufacturing, maintainable innovations, including sustainable energy frameworks, energy-proficient machines and gear, and waste-diminishment strategies, are fundamental to understanding the vision and objectives of maintainability inside the Industry 4.0 scene (Trappey et al., 2017). Maintaining IP rights for these revelations, from licenses and trademarks to exchange secrets, is essential for motivating ventures and proceeding with advancement. Yet, ensuring and imposing IP rights over maintainable innovations are more straightforward said than done. The rapidly moving innovation condition requests an equally fast IP-insurance methodology. In the meantime, they must put resources into comprehensive IP administration, which incorporates regular IP examining and consistently screens the challenge while utilizing the standard and criminal courses to authorize their IP rights when needed. Collaboration with IP workplaces and playing in worldwide IP treaties can build up the fitting obstructions for ecologically sound revelations (Cuellar et al., 2023).

8.3.3 Environmental regulations

Adherence to environmental laws is fundamental to sustainable Industry 4.0. These laws are designed to reduce the environmental harm from industrial activities and advance cleaner technologies. International agreements like the Paris Agreement have established aggressive targets for decreasing emissions and pressured companies to adopt sustainability measures. National and regional regulations reinforce these commitments. Examples of this are the Emissions Trading System (ETS) of the European Union (EU), where a cap is put on the total emissions from the industries, and a special edition of these subsidies allows companies to trade emissions allowances. The system encourages firms to reduce emissions by using cleaner-burning technologies and processes. Environmental regulation non-compliance comes with huge fines, legal liabilities, and a damaged reputation (Javaid et al., 2022).

8.3.4 Effects of rules on manufacturing process

Manufacturing industries are generally driven to follow best environmental practices as environmental regulations dictate manufacturing processes. Those requirements often mean compliance demands massive up-front investments in new equipment and infrastructure – investment in energy-efficient machinery, waste management systems, renewable energy power, etc.

To tackle these challenges, businesses must identify issues related to regulatory compliance as they arise and take steps to address them – all of which can be accomplished through a comprehensive environmental impact assessment (EIA). Operating with the ISO 14001 certification, a valid environmental management system, can further support compliance and improve general environmental performance.

The lifecycle approach, considering impacts at all stages, from raw material extraction to disposal, is among the essential factors to examine. This methodology allows for keen observation of possibilities to improve the environment, ensuring compliance with air-quality standards. Incorporating digital twins – virtual replicas of physical systems – can also improve efficiency in resource consumption, reduce waste, and increase sustainability (Guo et al., 2021).

This means that sustainable and just Industry 4.0 will need new laws and deserves consideration when considering the future of law. To validate that they are legally producing and promoting their products, businesses must continually invest in legal expertise and establish full compliance schemes per new rules and guidelines. Key considerations include:

1. Data Governance: Deploy a solid data governance framework to ensure adherence to data protection regulations. These measures range from compliance audits to staff training to new cybersecurity safeguards (Yebenes Serrano & Zorrilla, 2021).

2. IP Management: Design IP agreements to reference and structure cooperation projects to broadly protect IP rights. Monitoring and enforcing IP rights routinely to protect innovations (Trequattrini et al., 2022).
3. Environmental Compliance: This involves implementing an Environmental Impact Assessment and adopting favorable environmental and management practices. Be sure to keep yourself updated with changing regulations and invest in sustainable technologies to comply in the long run (Chiarini, 2021).
4. Risk Management: Identifying and managing legal risks through comprehensive risk assessments and compliance programs. Leverage legal technology products to simplify compliance procedures and to enhance the efficiency (Tupa et al., 2017).
5. Policy Engagement: Work with governmental and industry bodies to shape legislation underpinning sustainable Industry 4.0. Engage in public consultations and forums that can increase visibility and the ability to influence positive regulations (Kuo et al., 2019).

The regulatory landscape for businesses pursuing sustainable Industry 4.0 initiatives is a veritable minefield, and the legal challenges are complex and multi-layered. At the same time, data privacy and security, IP issues, and environmental regulation are all significant challenges to overcome through proactive strategy and strong compliance frameworks, including implementing sustainability into their operational best practices. Arming themselves with knowledge, legal guidance, and processes can save the business from a legal headache, help harness innovation, and, in turn, build a better business and sustainable future. The article presents some of the main legal and strategic challenges and essentials to be considered by any industry that seeks to effectively navigate the path of Industry 4.0 sustainably.

8.4 CASE STUDIES: NAVIGATING LEGAL CHALLENGES IN SUSTAINABLE INDUSTRY 4.0

If we take a closer look at some companies that successfully managed to face legal challenges in making their industry 4.0 sustainable, real-life cases show us the best practices and efficient strategies. This part shows a case-study analysis of companies such as Siemens, Tesla, and Unilever, the rigorous legal issues they confronted, and the solutions they followed.

8.4.1 Siemens

Company Background: Siemens is the global powerhouse in manufacturing and industrial digitalization with Industry 4. The company is working on smart manufacturing, IoT, and AI, focusing on sustainability. Legal Issues Faced Similarly, Siemens was met with significant difficulties when it began dealing with the implications of compliance with data privacy regulations such as the General Data Protection Regulation (GDPR). Data security and privacy were issues due to the volumes of data generated by connected devices and digital twins. Solutions Implemented 1. Siemens has implemented a robust data governance framework with strict data handling guidelines, data encryption, and user access controls for better GDPR compliance. 2. Regular Audits and Assessments: The company conducted frequent data protection impact assessments (DPIAs) to uncover and reduce privacy risks. Taking a proactive stance helped her maintain compliance with changing data privacy laws. 3. Siemens implemented extensive employee training in data privacy best practices and GDPR requirements, fostering a culture of data protection within the organization. 4 Takeaways Regular audits or DPIAs by data protection officers lead to an opportunity to discover adverse risks and to continue compliance. Privacy Culture: Training and awareness programs on basic privacy concepts are essential to building a data security culture (Privacy@Siemens, 2024).

Tesla: Safeguarding IP in Collaborative Settings Company Background Using cutting-edge technology and sustainable practices.

Tesla has been leading the way for electric vehicles and renewable energy solutions and is one of the biggest examples of innovation from Industry 4.0. Legal Issues Faced In a related post from August 2016, I wrote that while it was unlikely Tesla would sue suppliers using its patents, those patents would likely keep coming – industry knowledge had indicated continued activity concerning its battery technology. All that throughput in the IP world may have been the reason. Solutions Implemented 1. Clean IP Agreements: Tesla penned clear IP agreements with those partners, defining everything from ownership of jointly developed technologies to how they could be used and monetized. These were key to ensuring no conflicts and mutual gain from the work. 2. Open Innovation: Tesla took an open innovation route by providing some of its patents to the public. This approach helped drive industry-wide advancements in sustainable technologies without losing its competitive edge through proprietary innovations. 3. Enforce-IP: The firm actively pursued its IP portfolio and litigated against infringers to protect its technological treasures. IP Rights: Clear and comprehensive IP agreements are uno numero in joint projects to avert conflicts and shield innovations. Open Innovation – Sharing some IP encourages collective advancements within an industry, helping to protect core proprietary technologies (Tesla Comments on EDPB Guidelines 1/2020 on Processing Personal Data in the Context of Connected Vehicles and Mobility Related Applications, 2024).

Unilever: Compliance with Environmental Regulations

Company Background: Unilever is one of the leading multinational consumer goods companies at the forefront of sustainability by including environmental considerations in its supply chain and manufacturing practices. Legal Issues Faced: Not only did Unilever deal with environmental rules that had become very strict – directing companies to reduce their carbon footprints and get up to speed on sustainability practices; faced with complex rules, ensuring environments were up to code yet still conducive to productivity proved daunting. Solutions Implemented 1. EMS, Unilever has ISO 14001 Certification EMS (Environmental Management Systems) to have a statistical way to see how they can manage their environmental responsibilities. The certification facilitated the arithmetic of international and national environmental obligations. 2. Sustainable Tech Investments: The company believes in energy-efficient technologies and renewable energy sources to reduce carbon footprints and meet regulatory mandates on emissions reductions. 3. Life cycle assessments: Unilever employs a life cycle approach to measuring its products' environmental footprint, from extraction of raw materials to disposal. This 360-degree perspective allowed the business to discover inefficiencies and become more sustainable. Best Practices & Lessons Learned – Systematic Management: For instance, implementing an EMS like ISO 14001 makes it easier to meet regulations, and it can help to integrate sustainability throughout your organization. Lifecycle Approach: By considering the impact of a product on an environment throughout the entire product lifecycle, it is possible to identify opportunities for reducing environmental footprints and ensuring compliance with regulation (Unilever Environmental Policy, 2022).

The case studies illustrate various tactics companies use to navigate the legal hurdles confronting sustainable Industry 4.0. The various lessons and best practices that can be learned from Siemens' approach to data privacy, Tesla's management of IP in collaborative environments, and Unilever's compliance with environmental regulations are priceless.

Key Takeaways: 1. Proactive Law Compliance: It is important to understand the law in advance, conduct periodic assessments, and implement strategies to manage legal risks more effectively. 2. Collaboration IP Strategies: Defined IP agreements and open innovation Collaboration IP rules of engagement between parties while protecting proprietary technologies. 3. Responsibility: Adopting systematic management of materials handling and a lifecycle approach can improve sustainability and adherence to regulations. These examples offer the opportunity for sectors to be inspired and potentially navigate the challenges of embedding advanced technology and sustainability in these other sectors, support the establishment of compliance, and facilitate integration with areas of innovation.

8.5 FUTURE DIRECTIONS

Trends in the legal industry 4.0 and sustainability are changing every day. Many nascent legal trends are also likely to influence the 4.0 sustainability nexus. The first is the further focus on data ethics and responsible AI, as regulators and industries alike are faced with the consequences of decision-making affected by AI. Laws like the EU's AI Act strive to define pathways of responsible and ethical use of AI, and they could very well shape how companies create and utilize these technologies. Possible Evolutions of the Regulatory Frameworks: The whitepaper predicted that these current frameworks would naturally be subject to evolution to better accommodate the complexities of Industry 4.0.

On the other hand, it will ensure that international data privacy laws are more streamlined, making it easier for data to flow across borders while still protecting the privacy of personal data. The development of environmental, social, and governance indicators will likely result in more demanding environmental regulations, including greater emphasis on sustainability reporting and compliance initiatives. Regulatory sandboxes may also increase, which would facilitate the delivery of innovation through business while preserving control. Long-term consequences for industries and legal practices for a sustainable Industry 4.0 to become the rule of law, a long time will demand big changes in how companies and legal practices are developed. Businesses will be challenged to adapt to ever-shifting regulations and to implement best practices for compliance and risk management. This dynamic landscape will continue to fuel the need for legal experts in technology, sustainability, and international law. In addition, regulatory bodies need to work with companies to create flexible and adaptable legal frameworks that will balance the need for innovation and responsible practice.

8.6 CONCLUSION

To sum it up, the legally regulated field of sustainable Industry 4.0 is contributing challenges and opportunities equally. To do so, companies should focus on proactive legal compliance strategies, such as strong data governance, robust IP agreements, and an approach to environmental management systems. Risk management and policy engagement are crucial in mitigating legal risk to create a supportive regulatory environment. As the industry moves to sustainable Industry 4.0, industries of the future will face changing legal issues and regulatory frameworks that will require them to be nimble and ahead of the curve. Industries are already on their way to adopting IEC capability to enable innovation, meet sustainability targets, and maintain compliance in a fast-evolving world.

REFERENCES

Azadi, M., Moghaddas, Z., Cheng, T. C. E., & Farzipoor Saen, R. (2023). Assessing the sustainability of cloud computing service providers for Industry 4.0: A state-of-the-art analytical approach. *International Journal of Production Research*, 61(12), 4196–4213. https://doi.org/10.1080/00207543.2021.1959666

Chiarini, A. (2021). Industry 4.0 technologies in the manufacturing sector: Are we sure they are all relevant for environmental performance? *Business Strategy and the Environment*, 30(7), 3194–3207. https://doi.org/10.1002/bse.2797

Cuellar, S., Grisales, S., & Castaneda, D. I. (2023). Constructing tomorrow: A multifaceted exploration of Industry 4.0 scientific, patents, and market trend. *Automation in Construction,* 156, 105113. https://doi.org/10.1016/j.autcon.2023.105113

Dauvergne, P., & Lister, J. (2013). *Eco-business: A Big-Brand Takeover of Sustainability*. MIT Press.

Ferrari, A. M., Volpi, L., Settembre-Blundo, D., & García-Muiña, F. E. (2021). Dynamic life cycle assessment (LCA) integrating life cycle inventory (LCI) and Enterprise resource planning (ERP) in an industry 4.0 environment. *Journal of Cleaner Production,* 286, 125314. https://doi.org/10.1016/j.jclepro.2020.125314

Ghobakhloo, M. (2020). Industry 4.0, digitization, and opportunities for sustainability. *Journal of Cleaner Production,* 252, 119869. https://doi.org/10.1016/j.jclepro.2019.119869

Gokalp, M. O., Kayabay, K., Akyol, M. A., Eren, P. E., & Kocyigit, A. (2016). Big data for Industry 4.0: A conceptual framework. In *2016 International Conference on Computational Science and Computational Intelligence (CSCI)*, pp. 431–434. https://doi.org/10.1109/CSCI.2016.0088

Guo, D., Li, M., Lyu, Z., Kang, K., Wu, W., Zhong, R. Y., & Huang, G. Q. (2021a). Synchroperation in industry 4.0 manufacturing. *International Journal of Production Economics,* 238, 108171. https://doi.org/10.1016/j.ijpe.2021.108171

Hadi, K. (2024). *Intellectual property rights in the era of Industrial Revolution 4.0 in the perspective of legal protection. Proceeding International Conference on Law, Economy, Social and Sharia (ICLESS)*, 2, 16–38.

Javaid, M., Haleem, A., Singh, R. P., Suman, R., & Gonzalez, E. S. (2022). Understanding the adoption of Industry 4.0 technologies in improving environmental sustainability. *Sustainable Operations and Computers,* 3, 203–217. https://doi.org/10.1016/j.susoc.2022.01.008

Jazdi, N. (2014). Cyber physical systems in the context of Industry 4.0. In *2014 IEEE International Conference on Automation, Quality and Testing, Robotics*, pp. 1–4. https://doi.org/10.1109/AQTR.2014.6857843

Jope, A. (2022). *Unilever Environmental Policy*. www.unilever.com/files/f71d2b60-988c-4f8e-8626-c88f0a05fda4/unilever-environmental-policy.pdf

Khan, I. H., & Javaid, Mohd. (2022). Role of Internet of Things (IoT) in adoption of Industry 4.0. *Journal of Industrial Integration and Management*, 07(04), 515–533. https://doi.org/10.1142/S2424862221500068

Khan, M., Wu, X., Xu, X., & Dou, W. (2017). Big data challenges and opportunities in the hype of Industry 4.0. In *2017 IEEE International Conference on Communications* (ICC), pp. 1–6. https://doi.org/10.1109/ICC.2017.7996801

Kovács, G., Benotsmane, R., & Dudás, L. (2018). The concept of autonomous systems in Industry 4.0. *Advanced Logistic Systems – Theory and Practice*, 12(1), 77–87. https://doi.org/10.32971/als.2019.006

Kuo, C.-C., Shyu, J. Z., & Ding, K. (2019). Industrial revitalization via industry 4.0 – A comparative policy analysis among China, Germany and the USA. *Global Transitions,* 1, 3–14. https://doi.org/10.1016/j.glt.2018.12.001

Lasi, H., Fettke, P., Kemper, H.-G., Feld, T., & Hoffmann, M. (2014). Industry 4.0. *Business & Information Systems Engineering*, 6(4), 239–242. https://doi.org/10.1007/s12599-014-0334-4

Ocieczek, W., & Gajdzik, B. (2019). Social responsibility of business in Industry 4.0. *Zeszyty Naukowe Wyższej Szkoły Humanitas Zarządzanie*, 20(3), 89–102. https://doi.org/10.5604/01.3001.0013.7242

Paez, M., & Tobitsch, K. (2017). The industrial internet of things: Risks, liabilities, and emerging legal issues. *New York Law School Law Review*, 62, 217.

Penna, C. C. R., & Geels, F. W. (2012). Multi-dimensional struggles in the greening of industry: A dialectic issue lifecycle model and case study. *Technological Forecasting and Social Change*, 79(6), 999–1020. https://doi.org/10.1016/j.techfore.2011.09.006

Pinheiro, P., Putnik, G. D., Castro, A., Castro, H., Dal, B., & Romero, F. (2019). *Industry 4.0 and Industrial Revolutions: An Assessment Based on Complexity*. https://10.5937/fmet1904831P

Privacy@Siemens. (2024). siemens.com Global Website. www.siemens.com/global/en/company/about/compliance/dataprivacy.html

Rahanu, H., Georgiadou, E., Siakas, K., Ross, M., & Berki, E. (2021). Ethical Issues Invoked by Industry 4.0. In Yilmaz, M., Clarke, P., Messnarz, R., & Reiner, M. (Eds.), *Systems, Software and Services Process Improvement. EuroSPI 2021. Communications in Computer and Information Science* (pp. 589–606, vol. 1442). Springer, Cham. https://doi.org/10.1007/978-3-030-85521-5_39. https://doi.org/10.1007/978-3-030-85521-5_39

Rai, R., Tiwari, M. K., Ivanov, D., & Dolgui, A. (2021). Machine learning in manufacturing and industry 4.0 applications. *International Journal of Production Research*, 59(16), 4773–4778. https://doi.org/10.1080/00207543.2021.1956675

Rustad, M. L., & Koenig, T. H. (2019). Towards a global data privacy standard. *Florida Law Review*, 71, 365.

Schwartz, P. M. (2019). Global data privacy: The EU way. *New York University Law Review*, 94, 771.

Soares, M. N., & Kauffman, M. E. (2018). Industry 4.0: Horizontal integration and intellectual property law strategies in England. *Revista Opinião Jurídica (Fortaleza)*, 16(23), 268–289.

Soni, G., Kumar, S., Mahto, R. V., Mangla, S. K., Mittal, M. L., & Lim, W. M. (2022). A decision-making framework for Industry 4.0 technology implementation: The case of FinTech and sustainable supply chain finance for SMEs. *Technological Forecasting and Social Change*, 180, 121686. https://doi.org/10.1016/j.techfore.2022.121686

Sony, M., & Naik, S. (2019). Key ingredients for evaluating Industry 4.0 readiness for organizations: A literature review. *Benchmarking: An International Journal*, 27(7), 2213–2232. https://doi.org/10.1108/BIJ-09-2018-0284

Tesla, Inc. (2020). *Tesla comments on EDPB Guidelines 1/2020 on processing personal data in the context of connected vehicles and mobility related applications*. Brussels. Retrieved December 22, 2024, from www.edpb.europa.eu/sites/default/files/webform/public_consultation_reply/tesla_comments_edpb_guidelines_30apr2020_0_0.pdf

Trappey, A. J. C., Trappey, C. V., Hareesh Govindarajan, U., Chuang, A. C., & Sun, J. J. (2017). A review of essential standards and patent landscapes for the Internet of Things: A key enabler for Industry 4.0. *Advanced Engineering Informatics*, 33, 208–229. https://doi.org/10.1016/j.aei.2016.11.007

Trequattrini, R., Lardo, A., Cuozzo, B., & Manfredi, S. (2022). Intangible assets management and digital transformation: Evidence from intellectual property rights-intensive industries. *Meditari Accountancy Research*, 30(4), 989–1006. https://doi.org/10.1108/MEDAR-03-2021-1216

Tupa, J., Simota, J., & Steiner, F. (2017). Aspects of risk management implementation for Industry 4.0. *Procedia Manufacturing*, 11, 1223–1230. https://doi.org/10.1016/j.promfg.2017.07.248

Yebenes Serrano, J., & Zorrilla, M. (2021). A data governance framework for Industry 4.0. *IEEE Latin America Transactions*, 19(12), 2130–2138. https://doi.org/10.1109/TLA.2021.9480156

Chapter 9

Industry 4.0 performance measurement using key performance indicators for effective digital transformation

Nishal Murali and Ram Prasad Krishnakumar

9.1 INTRODUCTION

In recent years, the convergence of digital technologies and industrial processes has ushered in a transformative era known as Industry 4.0. This fourth industrial revolution is characterized by the integration of smart technologies, automation, data exchange, and advanced analytics into manufacturing and other sectors [1]. At the heart of Industry 4.0 are a plethora of innovative tools and technologies that empower organizations to enhance efficiency, productivity, and competitiveness in an increasingly interconnected world.

Industry 4.0 tools encompass a diverse range of solutions, including Internet of Things (IoT), artificial intelligence (AI), robotics, additive manufacturing (AM), and advanced analytics, among others. These tools offer unprecedented capabilities to monitor, analyze, and optimize various aspects of production, supply chain, and operations. However, to effectively harness the potential of these technologies and drive continuous improvement, organizations must establish clear performance measurement metrics and key performance indicators (KPIs). KPIs serve as vital benchmarks for evaluating the success and impact of Industry 4.0 initiatives. They provide actionable insights into key areas such as lead time, resource utilization, risk assessment, and capacity optimization. By tracking KPIs, organizations can identify strengths, pinpoint areas for improvement, and make informed decisions to drive operational excellence and strategic growth.

Furthermore, the need for mapping KPIs to specific Industry 4.0 tools arises from the complexity and interconnections of modern manufacturing and business processes. Each tool plays a unique role in enhancing operational efficiency, mitigating risks, and unlocking new opportunities. Mapping KPIs to corresponding tools enables organizations to align their performance measurement efforts with strategic objectives, ensuring that investments in technology translate into tangible business outcomes [2]. In this context, this paper explores the landscape of Industry 4.0 tools, delves

DOI: 10.1201/9781003470861-9

into KPIs relevant to these technologies, and underscores the importance of mapping KPIs to specific tools for comprehensive performance measurement. By understanding the symbiotic relationship between Industry 4.0 tools and KPIs, organizations can optimize their digital transformation journey, drive continuous improvement, and stay ahead in today's rapidly evolving business environment.

9.2 I4.0 TECHNOLOGIES

9.2.1 Internet of Things

IoT devices monitor environmental conditions in real time, minimizing spoilage risks and guaranteeing product freshness. Proactive surveillance ensures products arrive in pristine condition, delighting consumers with consistent quality. IoT boosts supply chain resilience through real-time data collection from sensors, enabling proactive issue identification, route optimization, and predictive maintenance [3]. In manufacturing, IoT is used for predictive maintenance, asset tracking, inventory management, and real-time monitoring of production processes [4]. IoT enables enhanced visibility and control over operations, improves asset efficiency, reduces downtime through predictive maintenance, and facilitates data-driven decision-making.

9.2.2 Cyber-physical systems

Cyber-physical systems (CPSs) integrate physical and digital components, optimizing temperature control and ensuring product quality. Their collaborative framework fosters efficient supply chain operations, resulting in safer and more efficient supply chains [5]. AI can be applied to provide a CPS with intelligent behavior. In case of machine damage, the machine agent decides the recovery method, such as overcoming the damage, cooperating with other machines to assign the work to an appropriate machine, or rescheduling [6]. According to the literature, among the elements of supply chain resilience (SCR) that are supported by CPS are SC configuration, flexibility, visibility, collaboration, information sharing, robustness, and velocity [7]. CPSs are used for smart manufacturing, energy management, intelligent transportation systems, and healthcare monitoring. CPSs improve efficiency, reliability, and safety of operations; enable real-time monitoring and control; and facilitate adaptive and responsive systems.

9.2.3 Artificial intelligence

AI algorithms optimize routes and predict disruptions, not only reducing spoilage risks and operational costs but also enhancing supply chain resilience by enabling rapid adaptation to changing conditions. Proactive decision-making fosters sustainable practices, ensuring that the supply chain remains robust and responsive to disruptions. The findings of various studies indicate that AI techniques have a positive impact on KPIs [8]. According to a survey conducted in 279 during 2021, the information processing capabilities of AI can be exploited to develop SCRes for enhancing SC performance [9]. In particular, the results indicate the mediating role of collaboration and adaptive capacity between AI and SCRes. They also argue that increasing flexibility or building redundancy can improve the ability of companies to adapt to changes in case of disruption. According to the literature, AI techniques enhance various elements of SCRes, such as redundancy (adaptive capability), flexibility, visibility, collaboration, agility, robustness, knowledge management, and velocity [10]. AI and machine learning are used for predictive maintenance, quality control, demand forecasting, and autonomous decision-making.AI and machine learning technologies enable automation of repetitive tasks, improve accuracy and efficiency, enhance product quality, and optimize resource utilization.

9.2.4 Big data analytics

Big data analytics (BDA) offer insights into supply chain performance, guiding informed decision-making to minimize waste and enhance efficiency [11]. Resource optimization ensures products reach consumers sustainably. BDA provides analysis and categorization of massive volumes of data into useful information and knowledge, which can support the decision-making processes in organizations. According to the literature, the elements of SCRes that are improved through the adoption of BDA are numerous. In particular, BDA is reported to enhance SC configuration, redundancy, flexibility, visibility, collaboration, agility, situation awareness, and information sharing [1,12, 13]. BDA is used for predictive maintenance, demand forecasting, quality control, and optimizing supply chain operations. BDA enables companies to gain actionable insights, improve forecasting accuracy, optimize processes, and identify new business opportunities.

9.2.5 Robotics and automation

Robotics involves the design and use of robots for performing tasks traditionally done by humans, while automation refers to the use of technology to control and monitor processes. Robotics and automation (R&A) are used for assembly, material handling, packaging, and inspection in manufacturing and logistics [14]. R&A improve productivity, reduce labor costs, enhance product quality, increase safety, and enable flexible and agile manufacturing processes.

9.2.6 Additive manufacturing

AM enables on-demand production of temperature-sensitive components, reducing waste and promoting sustainability. Customization ensures products remain fresh throughout their journey [12]. AM provides flexibility in design and operation, since different components and products can be produced on the same production line, achieving the customization of products when this is demanded. Moreover, AM can reduce the number of SC layers and suppliers [5,15]. AM is used for prototyping, customization, tooling, and low-volume production across various industries. AM enables rapid prototyping, design customization, cost-effective production of complex geometries, and on-demand manufacturing, leading to reduced lead times and inventory costs [16].

9.2.7 Cloud computing

Cloud computing (CC) offers scalable infrastructure for data storage and collaboration, empowering workers to adapt to changing conditions. For assessment of organizational resilience potential in small and medium enterprises of the automotive SC, cloud technology is used for logistics management, database management, and forecasting and planning demands [17, 18]. In the literature, the impact of CC on the elements of SCRes is reported. In particular, CC can enhance flexibility [19], visibility, collaboration, agility, information sharing, and risk management [20,21], CC is used for data storage, computing, analytics, and software-as-a-service (SaaS) applications. CC provides scalability, flexibility, and cost-effectiveness; enables real-time data access and collaboration; enhances data security; and facilitates remote access to resources and applications.

9.2.8 Augmented reality

Augmented reality (AR) technologies enhance worker productivity and safety, ensuring products are handled with care. Precision and confidence preserve product quality throughout the supply chain journey phase, allowing the detection of flaws without the physical prototypes [22]. This improves flexibility, and the risk of damage is minimized. Another example is the application of AR in staff training and support as real-time information, which can be provided either in the working environment or remotely [23]. Therefore, collaboration among partners is facilitated. Concluding the impact of AR on the elements of SCRes, the literature review provides evidence that AR contributes to the enhancement of flexibility, collaboration, knowledge management, and velocity [5,24]. AR and VR are used for training, maintenance, design visualization, and remote collaboration in various industries. AR and VR enhance training effectiveness, improve maintenance efficiency, enable remote assistance, and support design review and visualization, leading to reduced errors and downtime.

9.2.9 Blockchain

Blockchain (BC) ensures transparency and traceability, fostering trust and confidence in supply chain operations. Immutable ledgers reassure consumers, ensuring product authenticity and quality [24,25]. The findings revealed that BC improves SCRes, reducing the recovery time, the cost of disruption, and the number of affected partners. The researchers found that if BC is poorly implemented, then short disruptions have a strong negative impact on SCRes. Similarly, firms' resilience performance relies on the implementation of BC internally and across their SCs. According to the literature, among the elements of SCRes that are improved through the adoption of BC are SC configuration, flexibility, visibility, collaboration, agility, information sharing, robustness, and risk management [17,21,24]. BC is used for supply chain traceability, smart contracts, digital identity, and secure transactions in finance and logistics. BC enhances transparency, security, and trust in transactions; reduces fraud and errors; streamlines processes; and enables new business models and decentralized applications.

9.2.10 Advanced human-machine interface

Advanced human-machine interface (HMI) interfaces provide intuitive and interactive interfaces for humans to interact with machines and systems [15]. Advanced HMIs are used in manufacturing, process control, and consumer electronics for monitoring, control, and visualization. Advanced HMIs improve user experience, increase productivity, reduce errors, and enable real-time monitoring and control of systems and processes [26].

9.3 IMPORTANCE OF KEY PERFORMANCE INDICATORS

KPIs play a pivotal role in assessing the effectiveness and impact of Industry 4.0 initiatives. Scholars emphasize the need for organizations to define and measure relevant KPIs to monitor progress, identify areas for improvement, and align strategic objectives with operational activities [27]. Literature underscores the importance of selecting KPIs that reflect the organization's goals, priorities, and operational context. By focusing on key metrics such as lead time, resource utilization, quality performance, and customer satisfaction, companies can drive continuous improvement and operational excellence [28]. Moreover, research highlights the role of KPIs in fostering data-driven decision-making, enhancing transparency, and facilitating performance benchmarking both internally and across industry peers [14].

9.3.1 Key performance indicators

KPIs are the metrics that the organizations should calculate on a regular basis to monitor and evaluate processes. KPIs are among the organizational strategy elements that are heavily impacted by Industry 4.0. Many researchers report that monitoring KPIs can help build resilient supply chains. The aim of this research is to investigate which Industry 4.0 technologies can have an impact on the KPIs that are used for industrial efficiency measurement. In order to achieve this objective, first the constituent elements of performance indicators need to be determined, and which of the Industry 4.0 technologies can improve each of the selected elements needs to be examined. The appropriate KPIs are identified, and correspondingly the specific Industry 4.0 technologies that influence the selected KPIs are matched.

KPI 1 – Lead Time: Lead time represents the duration from order placement to receipt [29]. It encompasses order processing, production, and transportation, impacting customer satisfaction and inventory management. Lead time is crucial for customer satisfaction, inventory management, and overall operational efficiency. Shorter lead times often lead to higher customer satisfaction and reduced inventory holding costs. Lead time is typically measured in hours, days, or weeks depending on the context of the process or product being evaluated.

KPI 2 – Time to Recovery: Time to recovery denotes the period needed for a supply chain to restore normal operations post-disruption. This includes issue identification, corrective action implementation, and returning to pre-disruption levels, crucial for minimizing disruptions' impact. Time to recovery is critical for minimizing production losses, maintaining customer satisfaction, and ensuring business continuity. Time to recovery is measured in hours or minutes and can be tracked for different types of disruptions, such as equipment breakdowns, IT outages, or supply chain disruptions [30].

KPI 3 – Order Cycle Time: Order cycle time signifies the duration from order initiation to fulfillment, including processing, production, and delivery. Streamlining order cycle time enhances customer responsiveness and supply chain efficiency. Order cycle time directly impacts customer satisfaction, inventory turnover, and cash flow. Shorter cycle times enable faster response to customer demands and reduce inventory holding costs. Order cycle time is measured in hours, days, or weeks depending on the complexity of the order fulfillment process and the industry.

KPI 4 – Resource Utilization: Resource utilization focuses on efficiently using available resources like manpower, equipment, and materials, optimizing productivity and cost-effectiveness. Optimizing resource utilization helps in reducing costs, improving productivity, and maximizing return on investment

(ROI). Low resource utilization rates indicate underutilized capacity or inefficient processes. Resource utilization can be calculated as the ratio of actual resource usage to available capacity, expressed as a percentage [31].

KPI 5 – Risk Assessment Frequency: Risk assessment frequency reflects how often supply chain risks are evaluated. Regular assessments enhance resilience by identifying vulnerabilities and opportunities for improvement. Regular risk assessment helps in identifying and mitigating potential risks to the business, including operational, financial, and strategic risks. Risk assessment frequency is typically measured in intervals such as quarterly, semi-annually, or annually, depending on the industry regulations and organizational needs.

KPI 6 – Virtualization: Virtualization creates digital representations for simulation, enabling scenario analysis and optimization without disrupting actual operations. Virtualization enables efficient resource allocation, scalability, and flexibility in IT infrastructure, leading to cost savings, improved performance, and easier management. Virtualization can be measured by the percentage of IT infrastructure virtualized or the number of virtual machines deployed compared to physical resources.

KPI 7 – Forecasting: Forecasting predicts future demand and trends, aiding inventory management and decision-making for efficient supply chain operations. Accurate forecasting helps in optimizing inventory levels, production schedules, and resource allocation, leading to improved efficiency and customer satisfaction. Forecasting accuracy can be measured by comparing predicted values with actual outcomes using metrics such as Mean Absolute Percentage Error (MAPE) or Root Mean Square Error (RMSE).

KPI 8 – Interoperability: Interoperability enables seamless communication between supply chain systems and stakeholders, enhancing efficiency and collaboration. Interoperability enables integration, collaboration, and data sharing across various platforms and technologies, enhancing efficiency and innovation. Interoperability can be assessed based on the degree of compatibility, data exchange standards, and ease of integration between systems.

KPI 9 – Real-Time Data: Real-time data provides immediate access to current supply chain information, facilitating timely decision-making and response to changes or disruptions. Real-time data enables timely insights, faster decision-making, and proactive management of operations, leading to improved responsiveness and performance. Real-time data availability can be measured by assessing the latency or delay in data capture, processing, and dissemination across different systems and processes.

Through extensive literature review, we have identified key insights related to Industry 4.0 technologies that enhance resilience in the supply chain by addressing KPIs as shown in Table 9.1. These technologies collectively

Table 9.1 Mapping of I4.0 technologies to KPI for effective performance measurement

Key Performance Indicators (KPIs) / *Industry 4.0 Technologies and Tools*	*Lead Time*	*Time to Recovery*	*Order Cycle Time*	*Resource Utilization*	*Risk Assessment Frequency*	*Virtualization*	*Forecasting*	*Interoperability*	*Real-Time Data*
Artificial Intelligence					x		x		
Internet of Things (Iot)	x								
Cyber-Physical Systems						x			x
Big Data Analytics		x					x		
Additive Manufacturing	x		x					x	
Cloud Computing						x			x
HMI	x								x
Augmented Reality	x			x					
Blockchain					x				x
Robotics and Automation	x			x				x	

improve supply chain management by increasing efficiency, reducing downtime, enhancing accuracy, and boosting responsiveness. They facilitate shorter lead times, faster recovery, better service rates, and improved inventory and forecasting processes, resulting in a more resilient and effective usage of industry 4.0 technologies. Each technology plays a distinct role in optimizing different aspects of the industrial process chain, contributing to an integrated approach that significantly enhances overall performance and resilience. The literature underscores the transformative impact of Industry 4.0 tools on modern manufacturing and business processes, as well as the critical importance of KPIs in driving performance measurement and management. By integrating Industry 4.0 tools with relevant KPIs, organizations can unlock new levels of operational efficiency, innovation, and competitiveness in today's digital age.

9.4 INTEGRATION OF INDUSTRY 4.0 TOOLS AND KPIS

Scholars advocate for the integration of Industry 4.0 tools with KPIs to ensure effective performance measurement and management. This entails mapping specific KPIs to corresponding technologies and processes to track their impact on organizational outcomes [21]. Research suggests that aligning KPIs with Industry 4.0 tools enables organizations to leverage technology investments more strategically, identify areas of inefficiency or underperformance, and drive targeted interventions for process optimization [32]. Furthermore, literature highlights the role of advanced analytics and predictive modeling in deriving actionable insights from KPI data, enabling organizations to anticipate future trends, mitigate risks, and capitalize on emerging opportunities in the era of Industry 4.0 [18,33].

1. Internet of Things:
   - Lead time: IoT can optimize lead times by providing real-time data on inventory levels, production progress, and supply chain logistics.
   - Real-Time Data: IoT sensors continuously collect data, providing real-time insights into production processes and supply chain operations.
2. Big Data and Analytics:
   - Time to Recovery: BDA can analyze historical data to predict failures and reduce the time needed to recover from unplanned downtime.
   - Forecasting: BDA can generate accurate forecasts by analyzing historical trends and current data, helping in demand forecasting and resource planning.
   - Real-Time Data: BDA platforms process large volumes of data in real time, enabling timely decision-making.

3. Artificial Intelligence and Machine Learning:
    - Risk Assessment Frequency: AI and ML algorithms can continuously assess risks by analyzing data patterns and identifying potential issues.
    - Forecasting: AI and ML techniques can improve forecasting accuracy by analyzing complex data sets and identifying hidden patterns.
    - Real-Time Data: AI and ML models can process real-time data streams to provide insights and predictions instantaneously.
4. Robotics and Automation:
    - Lead time: R&A can reduce lead times by streamlining production processes and minimizing manual interventions.
    - Resource Utilization: R&A technologies optimize resource utilization by automating repetitive tasks and minimizing idle time.
    - Capacity Utilization: R&A systems can maximize capacity utilization by optimizing production schedules and minimizing downtime.
5. Additive Manufacturing (3D Printing):
    - Lead time: AM can reduce lead times by enabling rapid prototyping and on-demand production.
    - Order Cycle Time: 3D printing can shorten the order cycle time by producing customized parts quickly and eliminating the need for tooling.
    - Capacity Utilization: AM can improve capacity utilization by enabling flexible production and reducing setup times.
6. Cyber-Physical Systems:
    - Virtualization: CPSs enable virtualization of production processes, allowing for simulation and optimization before physical implementation.
    - Interoperability: CPSs facilitate interoperability between different machines and systems, enabling seamless communication and integration.
    - Real-Time Data: CPSs provide real-time data from sensors and actuators, enabling monitoring and control of physical processes.
7. Augmented Reality (AR) and Virtual Reality (VR):
    - Lead time: AR and VR can reduce lead times by facilitating remote collaboration, training, and virtual prototyping.
    - Resource Utilization: AR and VR technologies can optimize resource utilization by providing contextual information and guidance to workers.
    - Real-Time Data: AR and VR systems can overlay real-time data onto physical environments, enhancing situational awareness and decision-making.
8. Cloud Computing:
    - Virtualization: CC enables virtualization of IT resources, allowing for scalable and flexible infrastructure provisioning.

- Interoperability: CC platforms support interoperability between different systems and applications, enabling seamless data exchange and integration.
- Real-Time Data: CC provides real-time access to data and analytics tools, enabling timely insights and decision-making.

9. Blockchain Technology:
   - Risk Assessment Frequency: BC technology enhances risk assessment by providing transparent and tamper-proof records of transactions and contracts.
   - Interoperability: BC platforms support interoperability between different parties and systems, enabling secure and trusted transactions.
   - Real-Time Data: BC networks provide real-time visibility into transactions and data updates, enhancing transparency and traceability.
10. Advanced Human-Machine Interface:
    - Lead time: Advanced human-machine interface (HMI) systems can reduce lead times by improving operator efficiency and minimizing manual errors.
    - Interoperability: Advanced HMI interfaces support interoperability with various machines and systems, enabling seamless interaction and control.
    - Real-Time Data: Advanced HMI displays provide real-time data visualization and analytics, enabling operators to make informed decisions quickly.

Through systematic literature review (SLR), a hierarchical model has been developed based on the practices and the measures for sustainability and resilience analysis in an industry adaption. The latest technology is shown in Fig. 2. The resilience levels of industries are evaluated on the basis of practices like lead time, cycle time, recovery time, and more as recognized in the literature survey. When the firm initiates toward using I4.0 technologies for measuring the resilience, the level of impact can be assessed based on the practices listed; thus, they are kept at the top of the hierarchical model. The KPIs that enable firms to enhance resiliency are listed at the bottom level of the model, and their performance measurement using I4.0 technologies is discussed below.

9.5 CONCLUSION AND FUTURE WORK

The findings of this study pertaining to the impact of I4.0 technologies on KPI offer valuable insights for academics and practitioners interested in enhancing resilience within the organization. The Industry 4.0 technologies identified as improving KPIs for supply chain resilience can be considered by companies across various industries aiming to bolster their supply chain resilience. The integration of Industry 4.0 tools with KPIs presents significant opportunities for organizations to drive operational excellence and strategic growth. By mapping specific KPIs to corresponding technologies and processes, companies can effectively measure the impact of digital initiatives, identify areas for improvement, and optimize performance across the value chain. Through our analysis, we have identified several key findings:

1. Industry 4.0 tools such as the IoT, AI, robotics, and AM offer transformative capabilities to enhance efficiency, productivity, and innovation across industries.
2. KPIs serve as vital benchmarks for evaluating the success and impact of Industry 4.0 initiatives, providing actionable insights into key areas such as lead time, resource utilization, and risk assessment.

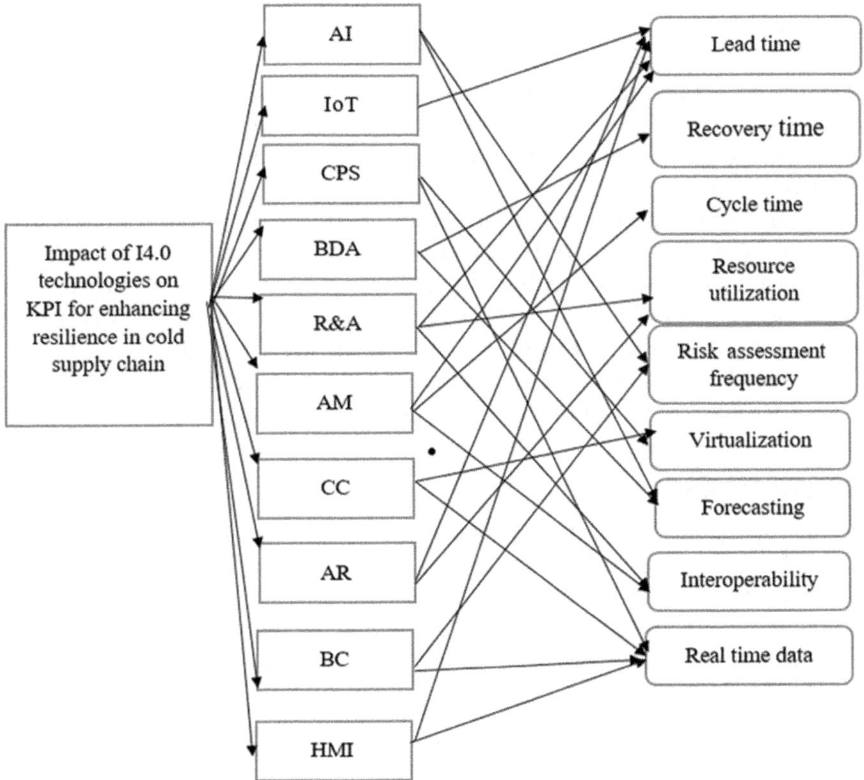

Figure 9.1 Hierarchical model for impact of I4.0 on KPI.

3. Mapping KPIs to specific Industry 4.0 tools enables organizations to align their performance measurement efforts with strategic objectives, ensuring that investments in technology translate into tangible business outcomes.
4. The integration of advanced analytics and predictive modeling enhances the value of KPIs by providing actionable insights and enabling proactive decision-making in real time.
5. Through practical examples and case studies, we have demonstrated how Industry 4.0 tools can be leveraged to optimize key performance metrics such as lead time, resource utilization, and capacity utilization.

Regarding limitations, it's important to acknowledge that our search strategy, like any SLR, might have missed or excluded some relevant references. This could include works not indexed in the selected databases or articles lacking specific keywords. However, we believe this is a minor limitation since the databases we used offer wide coverage in the field and are well regarded for their comprehensiveness. Future work will involve conducting empirical investigations to identify and evaluate KPIs for measuring resilience in organizations, utilizing decision support methods. Furthermore, the impact of Industry 4.0 technologies on these KPIs could be explored through case studies and expert interviews. The hierarchical model suggested is only a theoretical model based on literature, and to validate the model through consistency check, an analytical hierarchical process could be done to get a deeper insight about the interrelation between I4.0 technologies and resiliency practices and measures individually. In conclusion, the symbiotic relationship between Industry 4.0 tools and KPIs offers a pathway for organizations to unlock new levels of efficiency, innovation, and competitiveness in today's digital age. By embracing digital transformation and leveraging the power of data-driven insights, companies can position themselves for success in a rapidly evolving business landscape.

REFERENCES

1. Wang, S., Wan, J., Li, D., & Zhang, C. (2016). Implementing smart factory of Industrie 4.0: An outlook. *International Journal of Distributed Sensor Networks*, 12(1), 3159805.
2. Alam, T., & Biswas, P. (2019). A comprehensive survey on edge computing: Challenges and potential solutions. *Journal of Network and Computer Applications*, 141, 102673.
3. Xu, L. D., Xu, E. L., & Li, L. (2018). Industry 4.0: State of the art and future trends. *International Journal of Production Research*, 56(8), 2941–2962.
4. Belhadi, A., Kamble, S. S., Jabbour, C. J. C., Gunasekaran, A., Ndubisi, N. O., & Venkatesh, M. (2021). The role of big data analytics capabilities in bolstering

supply chain resilience: A COVID-19 perspective. *International Journal of Logistics Management,* 32(3), 893–921.

5. Schuh, G., Anderl, R., Gausemeier, J., Ten Hompel, M. and Wahlster, W., 2017. Industrie 4.0 maturity index. *Managing the digital transformation of companies, 61.*
6. Ivanov, D., & Dolgui, A. (2015). The digital supply chain twin: A simulation model. *IFAC-PapersOnLine*, 48(3), 1939–1944.
7. Zhang, Y., Qiu, M., Tsai, C. W., Hassan, M. M., & Alamri, A. (2017). Health-CPS: Healthcare cyber-physical system assisted by cloud and big data. *IEEE Systems Journal*, 11(1), 88–95.
8. Kusiak, A. (2018). Smart manufacturing must embrace big data. *Nature*, 544(7648), 23–25.
9. Grubert, E., & Guglielmo, F. (2019). The role of KPIs in Industry 4.0: An integrated approach. *IEEE Transactions on Engineering Management*, 66(3), 428–437.
10. Kagermann, H., Wahlster, W., & Helbig, J. (2013). Recommendations for implementing the strategic initiative INDUSTRIE 4.0: Securing the future of German manufacturing industry. *Final report of the Industrie 4.0 Working Group*, acatech – National Academy of Science and Engineering.
11. Yoo, S., Yoon, J., & Lee, D. (2021). Framework for assessing the impact of Industry 4.0 on manufacturing sustainability. *Sustainability*, 13(2), 945.
12. Kagermann, H., Wahlster, W., & Helbig, J. (2013). Recommendations for implementing the strategic initiative INDUSTRIE 4.0. *Final report of the Industrie 4.0 Working Group.*
13. Jiang, P., Wu, Q., & Xing, K. (2021). Big data analytics for sustainable smart manufacturing: A framework and case studies. *Robotics and Computer-Integrated Manufacturing*, 67, 102023.
14. Popescu, D., Popescu, A., & Pasat, G. (2017). Key performance indicators selection for the Industry 4.0 paradigm. *Management & Marketing. Challenges for the Knowledge Society*, 12(1), 139–147.
15. Rodrigues, M. P., & Martel, S. (2020). Key performance indicators for Industry 4.0. *Journal of Industrial Information Integration,* 17, 100125.
16. Popescu, D., Pânzaru, I., & Iancu, V. (2017). Key performance indicators for evaluating the Industry 4.0 implementation. *Proceedings of the International Conference on Business Excellence*, 11(1), 637–645.
17. Ghisolfi, V., Pereira, F. A., & de Freitas, G. R. (2017). Key performance indicators for Industry 4.0 in manufacturing operations: Literature review and perspectives. *Procedia Manufacturing,* 13, 1398–1405.
18. Roblek, V., Meško, M., & Krapež, A. (2016). A complex view of Industry 4.0. *SAGE Open*, 6(2), 2158244016653987.
19. Belhadi, A., Kamble, S. S., Zkik, K., Cherrafi, A., & Touriki, F. E. (2021). The integrated effect of Big Data Analytics, Lean Six Sigma and Green Manufacturing on the environmental performance of manufacturing companies: The case of North Africa. *Journal of Cleaner Production*, 312, 127789.
20. Grubert, E., & Brandt, A. R. (2019). Three considerations for modeling the social cost of carbon in integrated assessment models. *Climatic Change*, 154(3), 351–364.

21. Chang, S. I., Yen, D. C., Ng, C. S., & Kao, L. L. (2018). Industry 4.0 technologies and their applications in manufacturing. *International Journal of Production Research*, 56(18), 5779–5791.
22. Lasi, H., Fettke, P., Kemper, H. G., Feld, T., & Hoffmann, M. (2014). Industry 4.0. *Business & Information Systems Engineering*, 6(4), 239–242.
23. Truskolaski, S., Truskolaski, A., & Vasilev, D. (2020). Analyzing the relationship between Industry 4.0 technologies and key performance indicators. *International Journal of Production Economics*, 233, 107941.
24. Lu, Y. (2017). Industry 4.0: A survey on technologies, applications and open research issues. *Journal of Industrial Information Integration*, 6, 1–10.
25. Rojko, A. (2017). Industry 4.0 concept: Background and overview. *International Journal of Interactive Mobile Technologies (iJIM)*, 11(5), 77–90.
26. Jiang, Z., Cai, L., & Du, J. (2021). Leveraging artificial intelligence and machine learning for improved supply chain resilience. *Computers & Industrial Engineering*, 155, 107143.
27. Ivanov, D., & Dolgui, A. (2020). A digital supply chain twin for managing the disruption risks and resilience in the era of Industry 4.0. *Production Planning & Control*, 31(5), 377–393.
28. Hermann, M., Pentek, T., & Otto, B. (2016). Design principles for Industrie 4.0 scenarios. In *2016 49th Hawaii International Conference on System Sciences (HICSS)*, 3928–3937.
29. Wei, H., & Zhang, Z. (2021). Big data analytics and supply chain management: Research directions. *Operations Research Perspectives*, 8, 100169.
30. Yoo, Y., Henfridsson, O., & Lyytinen, K. (2021). Research in the era of digitalization: Trends and opportunities. *MIS Quarterly*, 45(2), 729–739.
31. Zhang, Y., Zhao, J., & Wan, J. (2021). Industry 4.0 and its impact on key performance indicators: A comprehensive study. *Journal of Manufacturing Systems*, 58, 322–337.
32. Kusiak, A. (2018). Smart manufacturing. *International Journal of Production Research*, 56(1–2), 508–517.
33. Alam, S., Khan, M., & Zia, A. (2019). Aligning Industry 4.0 with KPIs: A strategic approach for performance measurement. *Journal of Industrial Engineering and Management*, 12(4), 645–662.

Chapter 10

Machine learning applications in inventory management

A case study

Prakash Kumar, Vrishabh Narendra Bhonde, Sonu Rajak, and Malolan Sundararaman

10.1 INTRODUCTION

Inventory management has always been a vital process in the business environment. Managing inventory ensures an uninterrupted flow of raw materials and finished goods in the supply chain operations such as manufacturing and procurement [1, 2]. The objective of inventory management is to make the production process smooth, lower the inventory cost, and get the advantage of quantity discounts. Over the years, a variety of inventory models have been used to manage ordering frequencies, decide how much raw material and finished goods to store, and determine the ideal level of inventories to maintain during the manufacturing process in order to provide customers with uninterrupted service without any kind of delivery delay [3, 4]. However, one of the most important components of an inventory model is to predict the backorders. This would assist in identifying the products that are running low or completely out of stock, enabling an organization to promptly request restocking from its suppliers. As backordering helps in protecting the customer base, it is very much beneficial to business organizations. But, if this backordering scenario is not handled very well, it will create a monetary pressure on the supply chain to procure, manufacture, and deliver the products to customers within the given stipulated time. It will also have an impact on the degree of customer satisfaction, revenue, share price, and potential customer loss for the impacted company [5, 6]. To satisfy backorders, the actions put enormous pressure on various stages of the supply chain, which may result in extra labor and manufacturing costs and associated delivery charges. Therefore, solving backorder is a crucial step, and the solution is precise backorder prediction. A reliable inventory system is also a solution to avoid and handle backorders. But, nowadays, as the customers' demand is highly uncertain, it is impacting traditional supply chains in many ways, such as incorrect demand forecasting and misclassification of back-ordered products [4]. Recently, some companies are predicting the backorders using machine learning techniques to overcome the various tangible and intangible costs.

DOI: 10.1201/9781003470861-10

10.1.1 Current inventory level

There are positive, zero, and negative values in the current inventory levels. Equation 10.1 provides the safety stock calculation, which is used to determine the inventory level. The most used service level in the retail industry is assumed to be 90%.

$$\text{Safety stock} = Z \times \sigma \times D_{avg} \tag{10.1}$$

where D_{avg} is average demand. "Z" is the desired service level, and "σ" represents the standard deviation of lead time. To calculate the safety stock, the "Z" value can be fetched from the normal distribution table, which is 1.28 for 90%.

From the current inventory value, the safety stock is subtracted, which gives us values in ranges of negative, positive, and zero. This represents the status of the current inventory level. The negative inventory level could be because of various reasons like accidental duplication of sales or if the same type of inventory is maintained at different warehouses or if sales are recorded from the wrong store and products are sent to the wrong warehouses. The zero level of inventory shows no physical stock.

10.1.2 Problem statement

A backorder is an order that cannot be fulfilled now because the product is temporarily out of stock or may be due to a lack of supply but can guarantee the delivery of products to the customers requested by a certain date as there would be replenishment of the products. In simple words, a backorder is an order with a delayed reaching date. Backorders may happen due to improper forecasting, poor supply chain, and inventory management. If a lot of items are going on backorder, then that is an indication that an organization is not handling the operations as per the plan. If customers experience backorder regularly, they may switch their loyalty to competitors. Therefore, it will affect company's revenue and market share price. One cannot simply increase the inventory level to avoid backorder as it will create a monetary pressure on the supply chain because of storage costs, which the company will include in the product price, and it might result in losing customers to competitors. Thus, if we can predict items that have chances of backordering, it can be optimized at different levels, and it will avoid unexpected burdens on supply chain of an organization. Thus, there is a need for a backorder prediction system.

Further, the chapter is organized as follows. Section 10.2 reviews the recent literature. The methodology has been presented in Section 10.3. Section 10.4 discusses the results, and Section 10.5 presents the conclusion and draws the future research directions.

10.2 LITERATURE REVIEW

Enough research work is done in this stage to have a clear understanding of backordering in the business environment. The impact and necessity of backordering on the organization are studied. Previous research articles and websites of various manufacturers, sellers, etc. are referred to in this step.

Papers on machine learning (ML)-based material backorder prediction in inventory management were reviewed by De Santis *et al.* [7]. This study presents a predictive model for the target variable's imbalanced class problem. To check the model, various techniques such as receiver operating curve (ROC) and precision-recall curve have been suggested. A model of "gradient tree boosting" is adopted in this research. The AUC score of the implemented model is found to be 0.94. It has been suggested from this paper that ensemble learning achieves a greater AUC score than a single classifier. Blagging, a combination of under-sampling and tree ensemble, is also suggested to classify the target variable correctly. Lawal and Akintola [8] reviewed the paper related to the backorder predictive model using recurrent neural network (RNN). The authors have proposed an RNN model for countering backorder scenarios in the inventory management of an organization. Sampling techniques such as SMOTE and ADASYN ("Adaptive Synthetic Sampling") have been suggested to counter the imbalanced dataset. They have also used the method of Min-max scaler for data pre-processing. The result showed that RNN with ADASYN had performed way better with an F1 score of 0.889. Islam and Amin [9] reviewed papers related to predicting the probable backorder scenario in the supply chain. The authors have used "distributed random forest" (DRF) and "gradient boosting machine" (GBM) algorithms to predict the backorder of the supply chain. They have used a five-level and a four-level metric to indicate the inventory level and lead time, respectively. The levels suggest the dependency of several features on one another. It has been understood from this research that inventory level, lead time, demand forecast, and sales are the main features for predicting the backorder.

Li [10] studied backorder prediction related to a beer manufacturing company. The author has used the method of dimensionality reduction. As the dataset used by them is highly dimensional. Hence, to represent it in a simple format, "principal component analysis" (PCA), "linear discriminant analysis" (LDA), and null space LDA (LDA) are suggested. Different machine learning models, such as "Naive Bayes," "Support Vector Machine" (SVM), and "*K* Nearest neighbour" (KNN), are used here to predict the backorder. To see the accuracy of the implemented models, the AUC method is used, and it has been suggested that the KNN method gives the AUC maximum of 0.73 that has been implemented. Shajalal *et al.* [11] used deep neural network for backorder prediction.

The dataset used to carry out this research is highly imbalanced. Various techniques, including minority class weight boosting, randomized under-sampling, SMOTE, and a hybrid approach combining SMOTE and under-sampling, have been proposed here to achieve dataset balance. Various techniques of data balancing have been used here, which are then fed to the DNN model, and the accuracy is checked by using the AUC of ROC. A combination of SMOTE and random under-sampling is fed to the DNN model, and accuracy came out to be the best of all other methods. Chumongkhon and Sharedalal [12] focused on one of the barriers to a better service level, which is the backordering scenario. They have explored the inventory data to identify patterns in them so that the organization could not face the scenario of backordering. The authors have used various classification algorithms, such as gradient boosting, SVM, neural networks (NN), and logistic regression. Model validation is done using the area under the ROC curve. It has been suggested here that the neural network and gradient boosting algorithms would give the highest precision AUC of 0.95 of all the other algorithms that have been used. Ntakolia *et al.* [13] studied material backorder prediction in inventory management. The authors mainly focused on developing an explanatory pipeline of machine learning to evaluate the impact of various features used in predicting the backorder using machine learning techniques. They have done a comparative evaluation over commonly used classifiers such as RM, KNN, balance blagging (BB), NN, and light gradient-boosting machine (GBM). It has been found that BB reaches an AUC of 0.95. For this model, it was suggested that the most important features that were impacting the backorders were inventory level, quantity in transit, short-term sales, and the demand forecast. Chawla *et al.* [14] presented SMOTE. The authors have also suggested using the combination of random over-sampling and random under-sampling to handle the imbalanced target variable. The methods have been evaluated using the ROC convex hull strategy.

10.3 METHODOLOGY

10.3.1 Data collection

The dataset is an integral part of any study or analysis. The dataset that has been collected for this project is of a brewery manufacturing company. The data are of several years, and the backordering status of a product is recorded, i.e., whether the product is backordered or not. It is to be noted here that though there are lakhs of unique SKUs, it does not represent the unique products. As data are collected for several products at different timeline, and the status has been recorded, SKUs are only considered to identify it. SKU here can be considered as a serial no.

10.3.1.1 Data description

The dataset used here contains a total of 18 columns with 2,98,521 entries. The dataset is highly imbalanced because it contains a very small number of entries that show the product is backordered as compared to the entries that suggest the product is not backordered. Only 2% of the data show products with backorder status as "yes" and 98% as "no." The dataset has several other features, which can be studied in Table 10.1.

10.3.1.2 Problems in raw data

Three major problems in raw data have been collected, which would create a problem in building the machine learning models accurately.

- Dataset contained missing values
- Dataset contains heavy outliers.
- Dataset is highly imbalanced, with a positive class of target variable being a minority.

The above problems are solved by taking necessary actions in exploratory data analysis (EDA).

Table 10.1 Data description

Feature	*Description*
National_inv	Current inventory level of the product
Lead_time	Production plus transit time in weeks
In_transit_qty	Quantity of product in transit in units
Forecast_3_month	Forecast sales for the next 3 months
Forecast_3_month	Forecast sales for the next 6 months
Forecast_3_month	Forecast sales for the next 9 months
Sales_1_month	Sales quantity for the last 1 month
Sales_3_month	Sales quantity for the last 3 months
Sales_6_month	Sales quantity for the last 6 months
Sales_9_month	Sales quantity for the last 9 months
Min_bank	Minimum recommended amount of stock
Potential_issue	Flag for any kind of source issue for the part identified
Pieces_past_due	Parts delayed from source
Perf_6_month_avg	Performance of source in the last 6 months
Perf_12_month_avg	Performance of source in the last 12 months
Local_bo_qty	No. of stock orders delayed
Went_on_backorder	Product that went on backorder

10.3.2 Exploratory data analysis

After importing the dataset in Jupyter Notebook, necessary libraries were imported and basic EDA was carried out.

10.3.2.1 Target variable distribution

The target variable is a binary classification. That is, we are predicting whether the product is going to be backordered or not based on several historical features, as discussed above. A code snippet to see the distribution of the target variable is in Figure 10.1.

Here, we see that the target variable is highly imbalanced. Thus, to overcome the issue, SMOTE oversampling with a random under-sampling method is used.

10.3.2.2 Missing value check

Missing values from the dataset can be checked using Figure 10.2.

It has been found that the "lead time" column has a total of 122 missing values. These values can be imputed by employing closed entries. The given dataset contains heavy outliers, which might create a problem while building the machine learning models. Now, to check every column of the dataset, univariate analysis is done, and based on the output, outliers are detected.

10.3.3 Univariate analysis

To explore every column of a dataset, univariate analysis is used. Univariate analysis looks at different ranges of values in a column as well as the central tendency of the values. It describes the pattern of each variable on its own.

```
noCount = df.went_on_backorder.value_counts()[0]
yesCount = df.went_on_backorder.value_counts() [1]
print("Went on backorder = ", yesCount)
print("Not Went on backorder = ",noCount)
```

```
Went on backorder =  6577
Not Went on backorder =  291944
```

Figure 10.1 Distribution of target variable.

```
: df.isna().sum()
```

```
: sku                0
  national_inv       0
  lead_time        122
```

Figure 10.2 Missing values.

The code snippet and the boxplot of the "national_inv" column are seen in Figure 10.3.

From the above plot, it is very much clear that the "national_column" has heavy outliers.

Hence, to detect outliers, the method of 1.5 × IQR is used. The upper bound and lower bound of the column are fixed as equations 10.2 and 10.3:

$$\text{lower\_bound\_national\_inv} = \text{Q1\_national\_inv} - 1.5 \times \text{IQR\_national\_inv} \tag{10.2}$$

$$\text{upper\_bound\_national\_inv} = \text{Q3\_national\_inv} + 1.5 \times \text{IQR\_national\_inv} \tag{10.3}$$

The datapoints that fall below the lower bound or beyond the upper bound are not considered while plotting the boxplots now. Code snippet for fixing the upper and lower bounds as well as the further output of the boxplot is as shown in Figure 10.4.

It is to be noted from Figure 10.5 that the plot of the "Yes" category is more right-skewed than that of the "No" category, which implies that

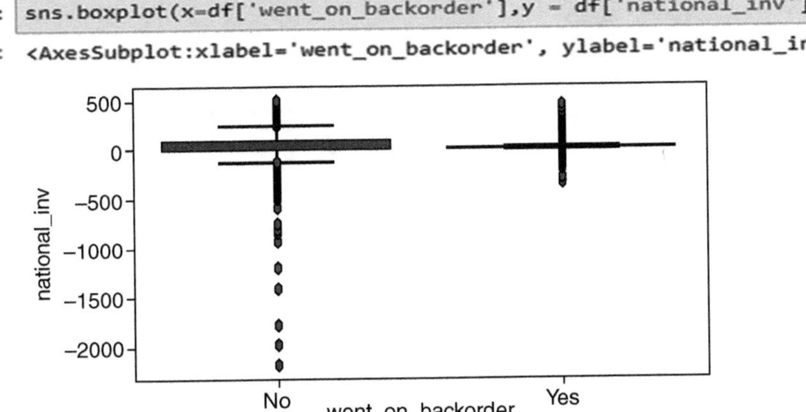

Figure 10.3 Boxplot of "national_inv" feature.

```
Q1_national_inv = df.national_inv.quantile(0.25) # Storing the 25th percentile value
Q3_national_inv = df.national_inv.quantile(0.75) # Storing the 75 th percentile value

IQR_national_inv = Q3_national_inv - Q1_national_inv  #Calculating the IQR

# Trying to detect outliers using this method where we define a new range
lower_bound_national_inv = Q1_national_inv - 1.5*IQR_national_inv
upper_bound_national_inv = Q3_national_inv + 1.5*IQR_national_inv
```

Figure 10.4 1.5 × IQR method code.

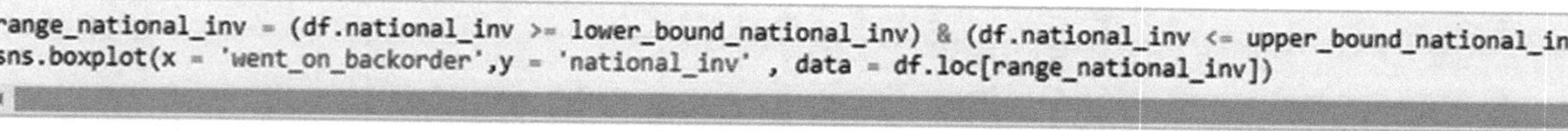

```
range_national_inv = (df.national_inv >= lower_bound_national_inv) & (df.national_inv <= upper_bound_national_inv)
sns.boxplot(x = 'went_on_backorder',y = 'national_inv' , data = df.loc[range_national_inv])
```

```
<AxesSubplot:xlabel='went_on_backorder', ylabel='national_inv'>
```

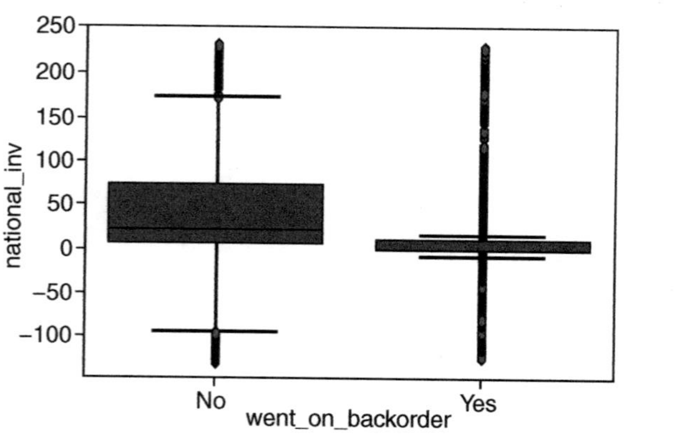

Figure 10.5 Boxplot after 1.5 × IQR.

there are more products that are not backordered. Also, most of the entries are concentrated at the lower end of the plot, and we also have negative inventory values. It can also be concluded from the boxplot that products with lower values of national inventory have more chance of going into backorder. The IQR is small, but the spread is high. Now, to remove the outliers, it is very much important to detect the outliers.

10.3.3.1 Outlier detection

Outliers are those data points that differ strongly in the sample of a population. To detect the outliers of the "national_inv" column, the percentile value of that column is found. The value from where there is an extreme switch is visible, that value is selected deep down so that very accurate outliers are detected. The code snippet for the same process is as shown in Figure 10.6. And, the code snippet to remove the outliers is shown in Figure 10.7.

Now, the next step is to see whether different features of the dataset are correlated with each other or not. Ideally, different features should be independent of each other to build the machine learning models.

```
for i in range(0,110,10):
    print(i,"percentile of values is ",df.national_inv.quantile(i*0.01))
print("-"*100)
print("Between 90-100 percentile")
for i in range(90,101,1):
    print(i,"percentile of values is ",df.national_inv.quantile(i*0.01))
print("-"*100)
print("Between 0-10 th percentile")
for i in range(0,10,1):
    print(i,"percentile of value is ",df.national_inv.quantile(i*0.01))
```

```
#print("1 percentile value is ",val[0])
for i in np.arange(1.0,0.0,-0.1):
    val = df['national_inv'].values
    val = np.sort(val,axis=None)
    print("{} percentile value is {} ".format(0+i,val[int(len(val)*(float(0+i)/100))]))
print("0 percentile value is ",val[0])
```

Figure 10.6 Percentile values.

```
df = df[(df.national_inv >= -74) & (df.in_transit_qty <= 721 )   &
  (df.forecast_3_month <= 1440.000) & (df.forecast_6_month <= 2541) &
  (df.forecast_9_month <= 3737) & (df.sales_1_month <= 580) & (df.sales_3_month <= 1637) &
  (df.sales_6_month <= 2889) & (df.sales_9_month <= 4151) & (df.min_bank <= 599) & (df.lead_time <= 20)
```

Figure 10.7 Outlier removal.

10.3.4 Pearson's coefficient of correlation

The linear correlation between two sets of data is measured by Pearson's correlation coefficient [15]. It is the ratio between the covariance of two variables and the product of their standard deviation. It has a value ranging between –1 and 1. As the value approaches 0, say from any end, we say that two variables are not related or independent of each other. If the coefficient is +1, two variables have a direct relationship and an inverse relationship for the value of –1. The output of Pearson's correlation code is shown in Figure 10.8. From Figure 10.8, it can be concluded that various sales are highly correlated with each other. Hence, only the "sales_1_month" column is considered for implementing machine learning models. Similarly, various forecasts are also correlated with each other, and hence, only "Forecast_3_month" is considered. "Lead-time," "national_inv," "In_transit_qty," and "potential_issue" columns are less correlated but affect the backordering, hence taken into consideration. Out of "Perf_6_month_avg" and "Perf_12_month_avg," only "Perf_6_month_avg" is considered, as both were

national_inv	0.0056	1	-0.0088	0.52	0.62	0.69	0.71	0.68	0.73	0.75	0.76	0.73	0.11	0.0067	0.013	-0.019
lead_time	0.0002	-0.0088	1	-0.06	-0.032	-0.049	-0.054	-0.061	-0.047	-0.03	-0.017	0.0033	0.0024	0.075	0.08	0.016
in_transit_qty	0.018	0.52	-0.06	1	0.56	0.61	0.62	0.64	0.66	0.65	0.64	0.58	0.11	0.016	0.014	0.096
forecast_3_month	0.024	0.62	-0.032	0.56	1	0.96	0.94	0.87	0.89	0.88	0.88	0.77	0.26	0.0047	0.005	0.15
forecast_6_month	0.02	0.69	-0.049	0.61	0.96	1	0.99	0.91	0.94	0.94	0.94	0.81	0.23	0.011	0.0099	0.13
forecast_9_month	0.02	0.71	-0.054	0.62	0.94	0.99	1	0.91	0.95	0.95	0.95	0.81	0.23	0.012	0.011	0.12
sales_1_month	0.0068	0.68	-0.061	0.64	0.87	0.91	0.91	1	0.96	0.94	0.93	0.8	0.2	0.015	0.015	0.18
sales_3_month	0.022	0.73	-0.047	0.66	0.89	0.94	0.95	0.96	1	0.98	0.97	0.82	0.21	0.014	0.014	0.15
sales_6_month	0.022	0.75	-0.03	0.65	0.88	0.94	0.95	0.94	0.98	1	0.99	0.82	0.21	0.015	0.013	0.13
sales_9_month	0.019	0.76	-0.017	0.64	0.88	0.94	0.95	0.93	0.97	0.99	1	0.83	0.21	0.013	0.011	0.12
min_bank	0.01	0.73	0.0033	0.58	0.77	0.81	0.81	0.8	0.82	0.82	0.83	1	0.18	0.011	0.017	0.1
pieces_past_due	0.12	0.11	0.0024	0.11	0.26	0.23	0.23	0.2	0.21	0.21	0.21	0.18	1	-0.0059	-0.0066	0.071
perf_6_month_avg	-0.0033	0.0067	0.075	0.016	0.0047	0.011	0.012	0.015	0.014	0.015	0.013	0.011	-0.0059	1	0.83	-0.008
perf_12_month_avg	-0.0032	0.013	0.08	0.014	0.005	0.0099	0.011	0.015	0.014	0.013	0.011	0.017	-0.0066	0.83	1	-0.0088
local_bo_qty	0.0078	-0.019	0.016	0.096	0.15	0.13	0.12	0.18	0.15	0.13	0.12	0.1	0.071	-0.008	-0.0088	1

Figure 10.8 Pearson's correlation coefficient.

correlated. The next step is to check the dependency of various numerical columns on the target variable. Ideally, there should be a dependency with the target variable. For this purpose, the "Point Biserial method" of statistics is used.

10.3.5 Point biserial method

The point biserial correlation coefficient is the correlation calculated between a continuous random variable (X) and a binary variable (Y).

Null Hypothesis: Variables are independent from each other.

Alternate Hypothesis: Variables are not independent from each other.

Alpha value is assumed to be 0.05. That is, at 95% confidence level, hypothesis is tested. The code snippet for Point Biserial hypothesis testing is shown in Figure 10.9.

After running the codes shown in Figure 10.9, we got the output as all the features are dependent on the target variable. So, by seeing Pearson's correlation output and Point Biserial output, several features are selected to build the model. Now, our dataset has one column named "potential_issue," which is categorical in nature. To find the relationship between this column and the target variable, a chi-square test is performed.

10.3.6 Chi-square test of independence

The chi-square test of independence is used because both the variables are categorical in nature.

Null Hypothesis: There is no correlation between the two variables.

```
numerical_features = ['national_inv', 'lead_time', 'in_transit_qty',
        'forecast_3_month', 'forecast_6_month', 'forecast_9_month',
        'sales_1_month', 'sales_3_month', 'sales_6_month', 'sales_9_month',
        'min_bank', 'pieces_past_due', 'perf_6_month_avg',
        'perf_12_month_avg', 'local_bo_qty']
```

```
#filling the Nan values using median imputation to find correlation
meanValue = copy_df['lead_time'].mean()
copy_df['lead_time'] = copy_df['lead_time'].fillna(meanValue)
```

```
target_variable="went_on_backorder"
```

```
alpha=0.05
for num in numerical_features:
  p_value = stats.pointbiserialr(copy_df['went_on_backorder'].values.ravel(),copy_df[num])[1]
  #print("p-value is :"+str(p_value))
  if p_value < alpha:
    print("p-value = "+str(p_value)+". There is a correlation between "+target_variable+" and "+num+". Reject Null Hypothesis i
  else:
    print("p-value = "+str(p_value)+". There is No correlation between "+target_variable+" and "+num+". Accept Null Hypothesis
```

Figure 10.9 Code snippet for point biserial hypothesis testing.

```
categorical_columns = ['potential_issue']
target_variable="went_on_backorder"
```

```
for cat in categorical_columns:
  table = pd.crosstab(df['went_on_backorder'],df[cat])
  stats,p,dof,expected = chi2_contingency(table)
  alpha=0.05
  print("p-value is "+str(p))
  if p <= alpha:
    print("There is correlation between "+target_variable+" and "+cat+". Reject the Null Hypothesis H0")
  else:
    print("There is no correlation between "+target_variable+" and "+cat+". Accept the Null Hypothesis H0")
```

Figure 10.10 Code snippet for chi-square test of independence.

Alternate Hypothesis: There is a correlation between the two variables.

Alpha value is assumed to be 0.05. That is, at 95% confidence level, hypothesis is tested. The code snippet for chi-square test of independence is shown in Figure 10.10. It has been found that the "Potential_issue" column and target variable are not independent. Thus, the "Potential_issue" column is also considered for model building. Now, the next step is data preprocessing, in which basic necessary steps have been taken based on the above EDA.

10.3.7 Data preprocessing

Preprocessing data is one of the crucial steps that are done before the actual model fits over the data. It creates a proper formatting of the data to feed to algorithms. There is a high chance that real-world data contain missing values, imbalanced dataset, and columns with zero more than 90%. This all should be cleaned from raw data, which is done in the preprocessing phase.

10.3.7.1 SMOTE

SMOTE is used to increase the minority class so that we can oversample the minority class. Using an oversampling technique called SMOTE, synthetic samples of a minority class are produced. Using interpolation between positive instances, it generates new instances based on the feature space.

10.3.7.2 Process of SMOTE

The total number of observations that are oversampled, "*N*," is displayed first. Usually, it is selected to ensure a 1:1 distribution of the binary class. A random positive class instance is chosen to begin each iteration. Next, the nearest neighbor of that instance is then determined by KNN (by default 5). Finally, a new synthetic instance is created based on N of these K classes. A distance metric is used to determine the gap between a vector and its neighbors. In order to add this difference to the preceding feature vector, the difference is multiplied by whatever random number lies between zero and one. Though this technique is quite useful, it creates noisy datapoints of one minority class. To overcome the effect, the majority class must be balanced with the minority class so that machine learning models could fit correctly. Hence, after SMOTE oversampling, random under-sampling method is also applied on the training dataset. The code to apply SMOTE and random under-sampling over the training dataset is shown in Figure 10.11.

Now, both the majority and minority classes are balanced, and we are ready to fit the model over the training dataset. We can also check it by plotting a graph of the minority and majority classes. Majority and minority data points of the target variable are shown in Figure 10.12.

```
from sklearn.datasets import make_classification
from imblearn.over_sampling import RandomOverSampler
from imblearn.under_sampling import RandomUnderSampler
from collections import Counter
```

```
# instantiating over and under sampler
os = SMOTE(sampling_strategy=0.5)
us = RandomUnderSampler(sampling_strategy=0.8)
# first performing oversampling to minority class
x_os, y_os = os.fit_resample(X,y_true)
print(f"Oversampled: {Counter(y_os)}")

# now to comine under sampling
x_us, y_us = us.fit_resample(x_os, y_os)
print(f"Combined Random Sampling: {Counter(y_us)}")
```

Figure 10.11 Code snippet to apply SMOTE with random under-sampling.

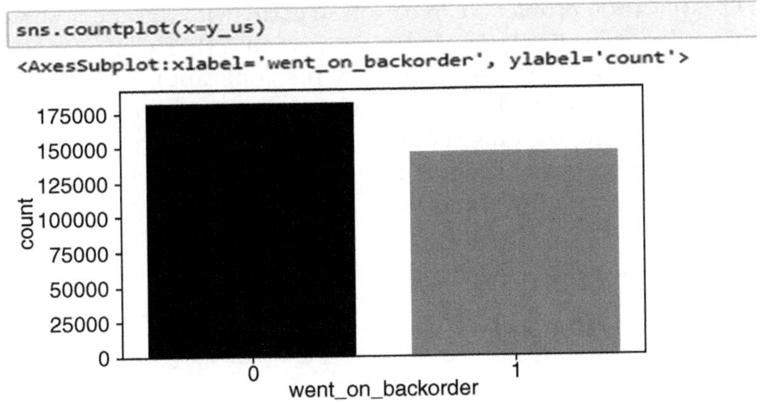

Figure 10.12 Plot of majority and minority datapoints of target variable.

```
X.describe()
```

	national_inv	lead_time	in_transit_qty	forecast_3_month	sales_1_month	potential_issue	perf_6_month_avg
count	296106.000000	296106.000000	296106.000000	296106.000000	296106.000000	296106.000000	296106.000000
mean	72.214673	6.828494	17.536055	72.681195	24.830456	0.001138	0.814491
std	100.806562	3.292333	47.833430	145.390948	48.299094	0.033717	0.221360
min	-74.000000	0.000000	0.000000	0.000000	0.000000	0.000000	0.000000
25%	6.000000	4.000000	0.000000	1.000000	1.000000	0.000000	0.740000
50%	26.000000	8.000000	0.000000	15.000000	6.000000	0.000000	0.890000
75%	96.000000	8.000000	11.000000	70.000000	24.000000	0.000000	0.980000
max	499.000000	20.000000	721.000000	1440.000000	580.000000	1.000000	1.000000

Figure 10.13 Clean dataset description to feed to ML models.

From Figure 10.12, we can see both the majority and minority classes are well balanced to each other. After all the above steps are completed, we can now check our dataset is much clean than the raw dataset as well as there are no outliers, no missing values, no misbalancing in the dataset, and only those features that are important and affecting our target variable are selected. The clean dataset that has been used to fit the various models can be seen in Figure 10.13.

Now, the final step of the methodology is fitting the various models over the clean dataset as seen in Figure 10.13.

10.3.8 Machine learning models

10.3.8.1 Logistic regression

Despite the name regression, the model is actually one of classification. It is an easy and effective way to solve linear and binary classification problems [16]. Classification problems can be solved easily with this method, which is widely used in industries [17]. Logistic regression (LR) can be conceptualized as linear regression with classification as the application [15]. There is one main difference between linear regression and logistic regression: logistic regression is limited to a range between 0 and 1 [18]. As well, logistic regression does not necessarily require a linear relationship between inputs and outputs. Equation 10.4 contains the LR equation.

$$P(x) = \frac{1}{1 + e\left(-\beta 0 + \beta 1 x\right)} \tag{10.4}$$

```
from sklearn.model_selection import cross_val_score
from sklearn.linear_model import LogisticRegression
lr = LogisticRegression()
lr.fit(X_train,y_train)
#lr_acc = cross_val_score(estimator=lr,X=train_x_os,y=train_y_os,cv=10)
```

Figure 10.14 Model-fitting code.

where β0 is the y-intercept of the equation and β1 is the slope of log-odds as a function of x.

The python code to fit the logistic regression over the given dataset is shown in Figure 10.14.

From the above code, we can see that logistic regression model is well fitted on the training dataset only. The confusion matrix and ROC curve of the fitted model are presented in Section 10.4.

10.3.8.2 Random forest

Random forests, also called random choice forests, are a type of ensemble learning technique that builds a large number of decision trees during training to solve problems like classification, regression, and other issues [18]. Its output is the classification selected by the largest number of trees in a random forest [17]. In a random forest classifier, multiple trees vote unit tickets for the most popular class, each constructed by sampling a random number separately from the input vector [19]. It uses bagging strategy that involves bootstrapping and aggregation.

10.3.8.3 XGBoost

A machine-learning method called gradient boosting is applied to classification and regression problems [19]. A decision tree is typically used as an ensemble of weak prediction models. The algorithm that results when a decision tree is the weak learner is known as a gradient-boosted tree. XGBOOST is short for eXtreme Gradient boosting package [20]. It is a scalable and effective use of the gradient boosting framework. In problems involving regression predictive modeling and classification, XGBoost outperforms structured or tabular datasets.

10.3.8.4 AdaBoost

AdaBoost or Adaptive boosting is an ensemble learning method. In order to improve weak classifiers, it iteratively learns from their mistakes. Despite the fact that a single classifier may not accurately predict a target variable's class, multiple weak classifiers can be used to build a strong classification model as each of them learns from the error of the others. This has been done using the AdaBoost ensemble learning technique.

10.4 RESULTS AND DISCUSSION

10.4.1 Exploratory data analysis results

1. Outliers from all numerical feature columns are identified and removed. A total of 2415 outlier rows are removed from the dataset to feed input to machine learning models.
2. Missing values from the "lead_time" column are imputed using the mean of nearest entries.
3. From Pearson's correlation matrix, forecasts, sales, and performance average features are highly correlated with each other. So, only the most recent and essential columns are used to forecast the target variable.
4. From the Point Biserial test to check the correlation between the target variable and numerical feature, it has been found that all the numerical features are impacting the target variable, as the p-value we got is less than alpha.
5. From the chi-square test of independence between the categorical feature "potential_issue" and target variable, it has been found that both features are not independent.
6. Columns with more than 90% zero have been dropped from the dataset.

10.4.2 Machine learning model results

Various machine learning models have been implemented, and the accuracy of the fitted models is calculated by using a confusion matrix and plotting the ROC curve to see the AUC. The confusion matrix summarizes the predicted results of a classification problem. Each class counts and summarizes the number of correct and incorrect predicted values. The confusion matrix and ROC curve are drawn on the test dataset.

10.4.2.1 Logistic regression

Python code for the confusion matrix and ROC plot is as shown in Figure 10.15.

Confusion matrix and ROC plot of logistic regression are as shown in Figures 10.16 and 10.17, respectively.

```
y_pred = lr.predict(X_test)
from sklearn.metrics import confusion_matrix,accuracy_score
cm = confusion_matrix(y_pred,y_test)
accuracy = accuracy_score(y_pred,y_test)
cm_display = ConfusionMatrixDisplay(confusion_matrix = cm,display_labels=[0,1])
cm_display.plot()
```

```
from sklearn import metrics
fpr, tpr,_ = metrics.roc_curve(y_test,y_pred)
auc = metrics.roc_auc_score(y_test,y_pred)
plt.plot(fpr,tpr, label = "AUC="+str(auc))
plt.ylabel("True Positive Rate")
plt.xlabel("False positive Rate")
plt.legend(loc=4)
plt.show()
```

Figure 10.15 Code snippet for confusion matrix and ROC.

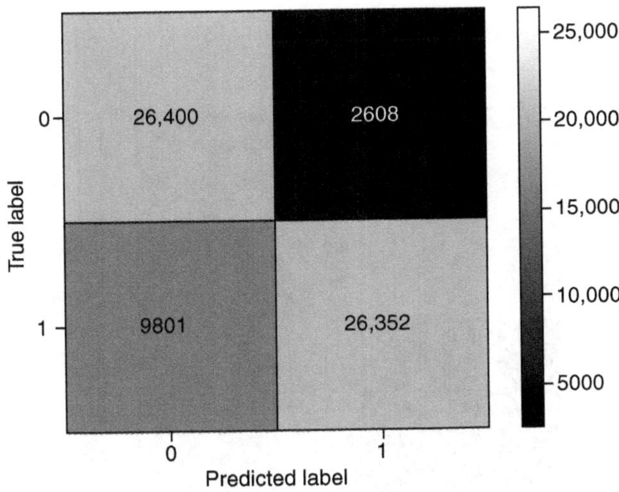

Figure 10.16 Confusion matrix for logistic regression.

From Figure 10.17, we can say that logistic regression predicts the target variable correctly with a precision of 81.96%.

10.4.2.2 Random forest

The same python code for plotting the ROC curve and confusion matrix is used as shown above; just the model inputs are changed. The confusion

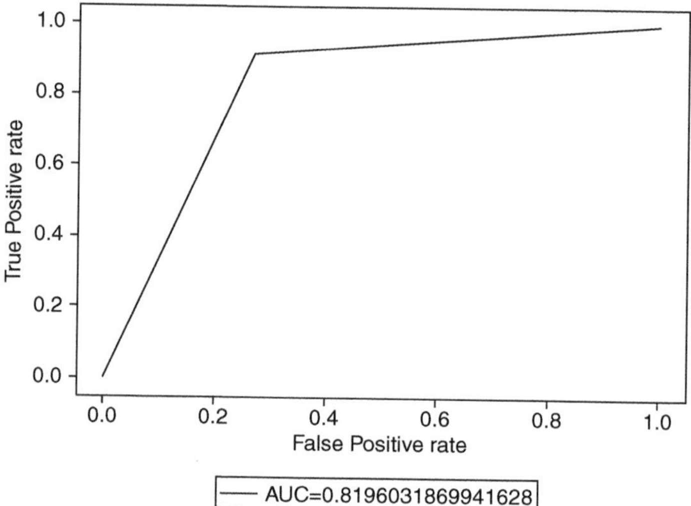

Figure 10.17 ROC for logistic regression.

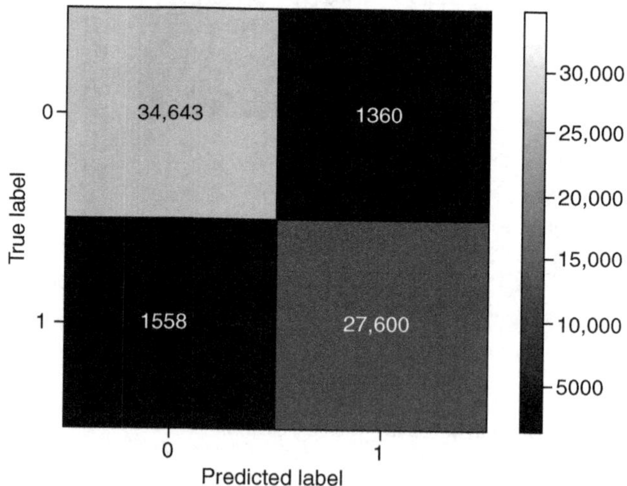

Figure 10.18 Confusion matrix for random forest.

matrix and ROC plot of random forests are shown in Figures 10.18 and 10.19, respectively.

From Figure 10.19 of the ROC plot, we can say that random forest is predicting the target variable correctly with a precision of 95.50%. From confusion it is seen that random forest is predicting the test dataset in a better way than that of logistic regression.

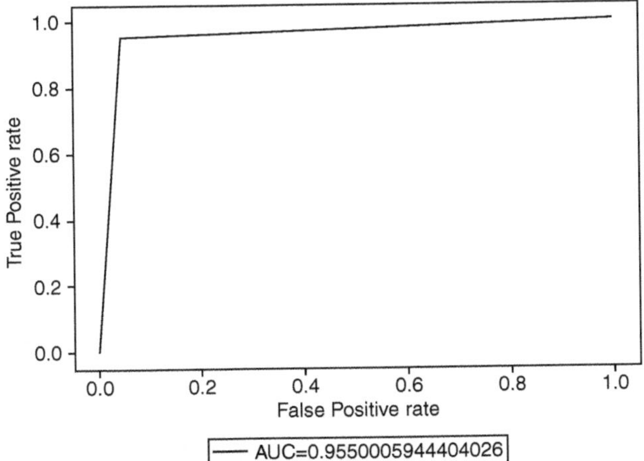

Figure 10.19 ROC for random forest.

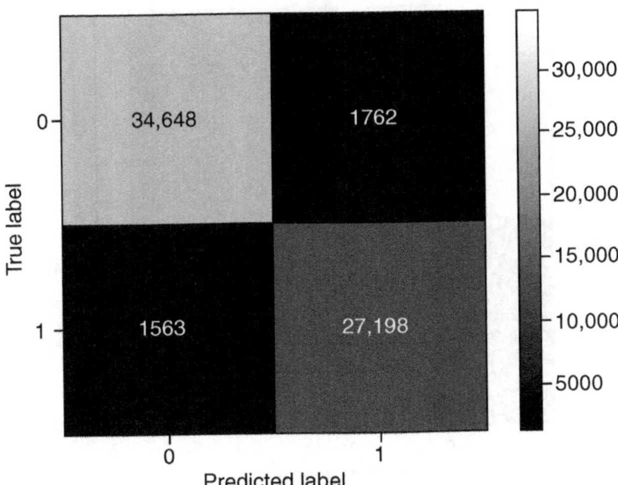

Figure 10.20 Confusion matrix for XGBoost.

10.4.2.3 XGBoost

The same python code for plotting the ROC curve and confusion matrix is used as shown above; just the model inputs are changed. The vonfusion matrix and ROC plot of XGBoost are as shown in Figures 10.20 and 10.21, respectively.

From Figure 10.21 of ROC plot, we can say that XGBoost is predicting the target variable correctly with a precision of 94.79%. From the confusion

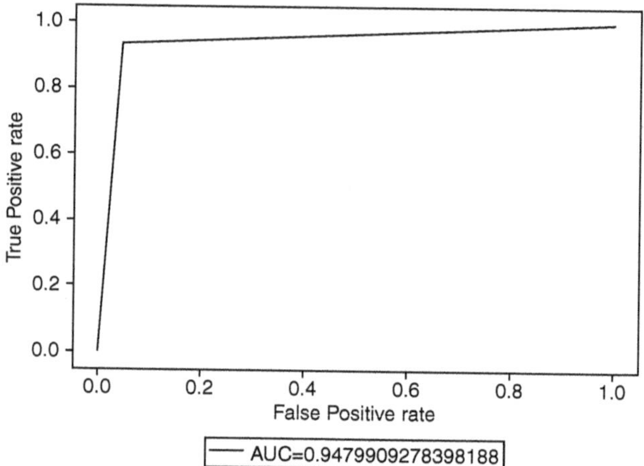

Figure 10.21 ROC for XGBoost.

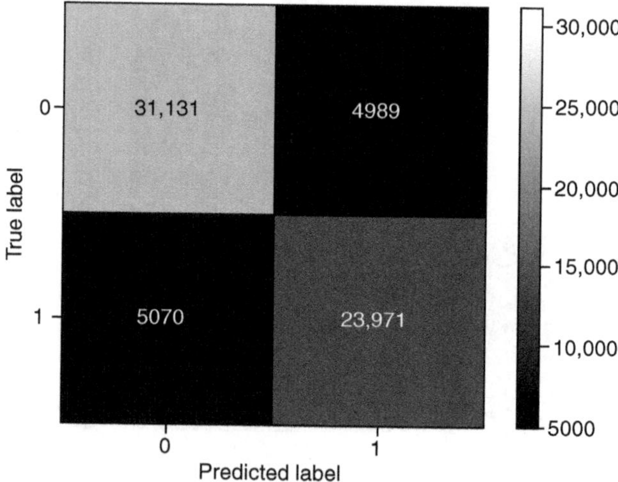

Figure 10.22 Confusion matrix for AdaBoost.

matrix, it is seen that XGBoost and random forest are predicting the test dataset almost with the same precision.

10.4.2.4 AdaBoost

The same python code for plotting the ROC curve and confusion matrix is used as shown above; just the model inputs are changed. The confusion matrix and ROC plot of AdaBoost are as shown in Figures 10.22 and 10.23, respectively.

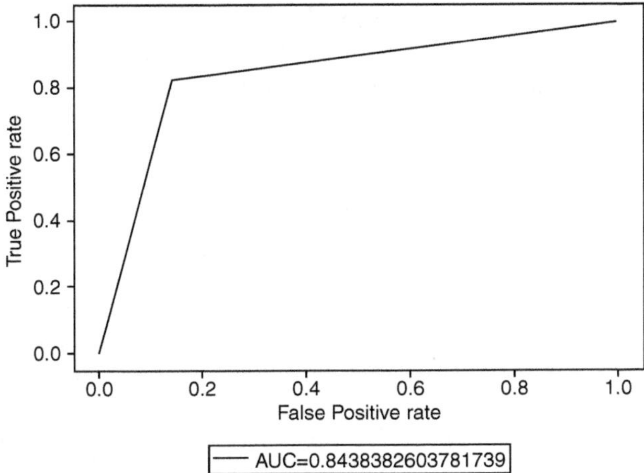

Figure 10.23 ROC for AdaBoost.

From Figure 10.23 of ROC plot, we can say that AdaBoost is predicting the target variable correctly with a precision of 84.38%. From the confusion matrix, it is seen that AdaBoost is predicting the target variable but with lesser precision as compared to random forest and XGBoost.

10.5 CONCLUSION

The following inferences can be drawn from the obtained results:

- After EDA and data preprocessing, we performed a test-train split of a dataset, and SMOTE with a random under-sampling technique was performed on the training dataset. Machine learning models are fitted after these techniques are applied, and the test dataset is predicted using various ML algorithms.
- From the ROC curve, we checked for area under the curve (AUC) and found out that random forest gives the best precision as compared to all the other models with an AUC of 95.50%.
- After random forest, XGBoost is also giving a precision of 94.79%, which is almost the same as the random forest model.

10.5.1 Scope for future work

- As a part of future work, we can have hyperparametric tuning of all the models that have been implemented so that we can have higher precision and can predict the products going to backorder correctly.

- Different models of deep learning can be the future scope to see if we get better accuracy of the models.
- Advanced sampling techniques other than SMOTE can also be used to get the balance of the imbalanced data.

REFERENCES

1. Dinh Canh, T. (2019). *Applying Machine Learning in Predicting Backorder: An Onpoint E-commerce Company Case Study* (Doctoral dissertation, International University-HCMC).
2. Rajak, S., Parthiban, P., & Dhanalakshmi, R. (2019). A hybrid metaheuristics approach for a multi-depot vehicle routing problem with simultaneous deliveries and pickups. *International Journal of Mathematics in Operational Research*, *15*(2), 197–210.
3. Rodger, J. A. (2014). Application of a fuzzy feasibility Bayesian probabilistic estimation of supply chain backorder aging, unfilled backorders, and customer wait time using stochastic simulation with Markov blankets. *Expert Systems with Applications*, *41*(16), 7005–7022.
4. Rajak, S., Parthiban, P., & Dhanalakshmi, R. (2018). Selection of transportation channels in closed-loop supply chain using meta-heuristic algorithm. *International Journal of Information Systems and Supply Chain Management (IJISSCM)*, *11*(3), 64–86.
5. Srivastav, N., & Srivastav, A. (2019, February). Solving an intractable stochastic partial backordering inventory problem using machine learning. In *International Conference on Emerging Technologies in Computer Engineering* (pp. 240–249). Springer.
6. Rajak, S., Parthiban, P., & Dhanalakshmi, R. (2021). A DEA model for evaluation of efficiency and effectiveness of sustainable transportation systems: A supply chain perspective. *International Journal of Logistics Systems and Management*, *40*(2), 220–241.
7. De Santis, R. B., de Aguiar, E. P., & Goliatt, L. (2017, November). Predicting material backorders in inventory management using machine learning. In *2017 IEEE Latin American Conference on Computational Intelligence (LA-CCI)* (pp. 1–6). IEEE.
8. Lawal, S., & Akintola, K. (2021). A product backorder predictive model using recurrent neural network. *IRE Journals*, *4*(8).
9. Islam, S., & Amin, S. H. (2020). Prediction of probable backorder scenarios in the supply chain using distributed random forest and gradient boosting machine learning techniques. *Journal of Big Data*, *7*(1), 1–22.
10. Li, Y. (2017). *Backorder Prediction Using Machine Learning for Danish Craft Beer Breweries*. Aalborg University.
11. Shajalal, M., Hajek, P., & Abedin, M. Z. (2023). Product backorder prediction using deep neural network on imbalanced data. *International Journal of Production Research*, *61*(1), 302–319. https://doi.org/10.1080/00207543.2021.1901153.

12. Chumongkhon, V., & Sharedalal, R. Predicting product backorder using machine learning algorithms, 1–4. https://bpb-us-e1.wpmucdn. com/sites.psu.edu/dist/b/86690/files/2018/04/Project_Report103vmtb.pdf.
13. Ntakolia, C., Kokkotis, C., Karlsson, P., & Moustakidis, S. (2021). An explainable machine learning model for material backorder prediction in inventory management. *Sensors*, *21*(23), 7926.
14. Chawla, N. V., Bowyer, K. W., Hall, L. O., & Kegelmeyer, W. P. (2002). SMOTE: Synthetic minority over-sampling technique. *Journal of Artificial Intelligence Research*, *16*, 321–357.
15. Breitenbach, J., Haileselassie, S., Schuerger, C., Werner, J., & Buettner, R. (2021, December). A systematic literature review of machine learning tools for supporting supply chain management in the manufacturing environment. In *2021 IEEE International Conference on Big Data (Big Data)* (pp. 2875–2883). IEEE.
16. Sharma, S., Deepika, D., & Singh, G. (2021, December). Intelligent warehouse stocking using machine learning. In *2021 IEEE International Conference on Mobile Networks and Wireless Communications (ICMNWC)* (pp. 1–6). IEEE.
17. Lolli, F., Balugani, E., Ishizaka, A., Gamberini, R., Rimini, B., & Regattieri, A. (2019). Machine learning for multi-criteria inventory classification applied to intermittent demand. *Production Planning & Control*, *30*(1), 76–89.
18. Ensafi, Y., Amin, S. H., Zhang, G., & Shah, B. (2022). Time-series forecasting of seasonal items sales using machine learning – A comparative analysis. *International Journal of Information Management Data Insights*, 2(1), 100058.
19. Sharma, G., & Patil, S. (2022). Extreme gradient boosting model-based forecasting of big data online sales record. *SAMRIDDHI: A Journal of Physical Sciences, Engineering and Technology*, *14*(01), 112–119.
20. Priore, P., Ponte, B., Rosillo, R., & de la Fuente, D. (2019). Applying machine learning to the dynamic selection of replenishment policies in fast-changing supply chain environments. *International Journal of Production Research*, *57*(11), 3663–3677.

Chapter 11

Research issues in Industry 4.0

Priyanka B. Gaikwad and Vishal Ashok Wankhede

11.1 INTRODUCTION

Industry 4.0, the Fourth Industrial Revolution, is the current trend of automation and data exchange in manufacturing technologies (Xu, Xu, and Li 2018). Its scope increases the level of automation, data exchange, and self-monitoring with the intelligent use of Internet of Things (IoT), artificial intelligence (AI), and cyber-physical systems (CPS). There is much evidence already showing how it has drastically improved efficiency, productivity, and customization. Industry 4.0 makes it possible for the interconnected, intelligent factory where machines, systems, and processes can communicate with and respond to one another autonomously (Tran, Nguyen, and Hoang 2019). Studies suggest that the blockchain may even disrupt industries beyond manufacturing, including logistics, health, and urban planning, radically changing traditional business models and economic architecture.

11.1.1 Importance of addressing research challenges

Based on the integration of CPS, IoT, AI, and big data analytics, Industry 4.0, known as the 4th industrial revolution, initiates a new stage in the evolution of manufacturing and production processes in general (Koh, Orzes, and Jia 2019; Wankhede and Vinodh 2022). While the new abilities offer great potential to improve productivity, efficiency, and innovation, these also bring in many research challenges that need to be solved (both to realize the good parts and to manage the resulting risks).

11.1.1.1 Boosting interoperability and harmony

Integrating dynamics with diverse systems and technology stacks is a difficult challenge in the grand Industry 4.0 scheme. In manufacturing settings, factories are now becoming intricate ecosystems where machines, devices, and systems from multiple manufacturers need to communicate and interoperate properly (Mourtzis, Angelopoulos, and Panopoulos 2022). To

DOI: 10.1201/9781003470861-11

overcome this hurdle, we need to focus on researching universal standards, protocols, and frameworks that enable easy integration. A higher level of interoperability allows data to effortlessly spread along the production pipeline, ensuring real-time decisions and process optimization for staying competitive in an increasingly fast market.

11.1.1.2 Operationalizing and exploring big data

Industry 4.0 has brought with it a proliferation of sensors and connected devices, which produce an enormous quantity of data (Majid et al. 2022). Another tremendous research challenge is efficiently handling and interpreting these data to get meaningful and actionable results. Existing legacy data management systems seem grossly outdated and ill-equipped to handle the massive amount of data that is being generated at an extremely high speed and in different formats inherent in operating smart factories. Further research on advanced data analytics tools and methods, which use machine learning algorithms and have real-time processing capabilities, is also warranted. Its tools also assist in predictive maintenance, quality control, and supply chain optimization, which can result in cost savings and improved operational efficiency.

11.1.1.3 Ensuring cybersecurity

In the Industry 4.0 scenario, with data and connectivity crossing, the system sacrifices the security. Significant research deals with cybersecurity, which is the main research area for protecting sensitive data and maintaining the integrity and reliability of the industrial control systems (Adeoye 2024). Studies in the realm of cryptography care largely about creating structures and algorithms that are strong against attacks – we have encryption, detection, and secure gossip. Addressing these cybersecurity challenges will enable organizations to prevent costly disruptions and protect their intellectual property, keeping the integrity of their customers and stakeholders intact.

11.1.1.4 Enhancing human-machine interaction

Human-centered design becomes even more important as automation and robotics become more pervasive (He et al. 2022). New user interfaces must be developed, and collaborative robots (cobots), for instance, will have to be taught so that people can work together with them safely and effectively in future factories. This should include ergonomics, cognitive load, and intuitive control systems. The benefit of a better human-machine collaborative system is to enable safer, more convenient, and efficient collaboration with human workers, to allocate the resource to human workers on more complex, creative tasks, and robots for repetitive or hazardous work.

11.1.1.5 Promoting sustainability

Through smart use of resources and waste minimization, Industry 4.0 technologies can cut environmental impact considerably. Still, the application of these technologies does create new sustainability problems (Javaid et al. 2022). It has never been more important to experiment to arrive at manufacturing solutions that are compatible with our long-term sustainability, energy efficiencies, and recyclable/recoverable facets of materials. Overcoming these challenges not only deals with becoming environmentally compliant but also resonates with increasing consumer push for eco-friendly products, elevating the market position of any brand.

11.1.1.6 Policy and regulatory environment

Industry 4.0 technologies develop so quickly that they can leave existing regulations in the dust. Policy and regulatory research are also needed to facilitate legal compliance in deploying new technologies and ethical issues. This will be related back to issues such as data privacy rights, intellectual property rights, and how automation affects decision-making in the labor market. Proactive policy research is necessary to shape appropriate regulations that are conducive to innovation while ensuring public interests.

11.1.1.7 Promoting innovation and economic competitiveness

In order to accelerate future research directions from Industry 4.0 research challenges, it is important to identify crucial issues that need to be researched for innovation and global competitiveness. Organizations that invest in R&D can take advantage of emerging technologies as they become available, adapt to changing markets, and address changing customer demands. Through the conquest of technological and operational challenges, businesses will be able to create and release innovative products that better serve their audience while at the same time optimizing their processes and becoming more efficient and profitable.

11.1.2 Impact on global industries and economies

Driven by the integration of digital technologies including CPS, the IoT, AI, and big data analytics, around Industry 4.0, the era of Industry 4.0 is transforming industries and economies globally (Wankhede and Vinodh 2022). At the heart of it all is the Fourth Industrial Revolution and the way it is altering manufacturing processes, supply chains, and even business models – with effects that are destined to be deeply felt in global industries and economies. At its core, Industry 4.0 has the potential to revolutionize manufacturing processes by the influence of smart factories and more advanced automation. Interconnectivity of CPS with IoT allows monitoring and controlling the manufacturing activities in real-time mode, which in turn increases the productivity of the system by minimizing downtime and the improvement in the product quality. State-of-the-art robotics and AI-controlled systems enable extreme accuracy and flexibility in production, which is necessary for mass customization and just-in-time manufacturing. This not only increases productivity but also decreases operational costs, thereby becoming the vanguard for manufacturers in the global market. Industry 4.0, enabled by emerging technologies such as IoT, AI, and big data, is transforming entire global industries and economies. Smart factories and automation add a more efficient, cost-effective, and qualitative dimension to manufacturing. Real-time tracking and predictive analytics optimize supply chains to remain in tune with the inventory and pace incoming data.

Additive manufacturing and AI-driven insights allow products to be developed and brought to market quickly, which speeds up the path of innovation and enables new market opportunities (Xiong et al. 2022). Digital skillset becomes the most sought-after skill, driving education and training expenses for upskilling the workforce (Vinodh and Wankhede 2021). From an economic perspective, Industry 4.0 raises productivity, profitability, and competitiveness by reducing the cost of doing business – pulling in foreign investment and driving economic expansion. Yet, areas that do not take advantage of these technologies get left, resulting in economic inequality. Environmentally, smart manufacturing and energy-efficient technologies conserve resources and decrease emissions, aligning with sustainability initiatives. Data protection, cybersecurity, and ethical concerns (ensuring data privacy, cybersecurity, and addressing AI ethical concerns): as such, the regulatory frameworks must modernize and cope with these challenges and poorly bridge the divide created by the digital transformation across the developed and developing regions, respectively.

11.2 CURRENT RESEARCH CHALLENGES

11.2.1 Integration and interoperability

11.2.1.1 Challenges in integrating diverse systems and technologies

Interoperability, standardization of communications protocols – one of the biggest challenges of integrating diverse systems and technologies in the context of Industry 4.0 is that there are no universal protocols and interoperability frameworks for those (Anthony Jnr 2024). Seamless integration with different vendors and legacies is, however, often difficult and expensive with the deployment of proprietary technologies. The goal of most current research is to develop universal standards and middleware approaches to bridge these gaps and allow different components to communicate and work effectively together, ultimately increasing the system's efficiency and functionality.

11.2.1.2 Standards and protocols

Setting global standards and protocols is vital so that Industry 4.0 technologies can be integrated without any hiccups. This research banks on coming up with shared standards that maintain consistency and cross-compatibility between a variety of platforms, devices, and such. Standardization efforts from international bodies like the International Organization for Standardization (ISO) and Industrial Internet Consortium (IIC) are crucial in creating more unified ecosystem, as they can foster interoperability between different technologies and bring improved global adoption and use of Industry 4.0 solutions.

11.2.2 Data management and analytics

11.2.2.1 Big data management

Industry 4.0 is built on data, and Industry 4.0 creates so much of this with all the different sensors and connected devices. Due to the huge volume, velocity, and variety of data, traditional data storage and processing systems cannot handle this data. We focus our research on building large-scale storage that can efficiently store big data and rapidly analyze big data in high-level languages using big data type systems on scalable storage systems like distributed databases and cloud platforms. Improved data handling capabilities are being achieved by advanced data compression and indexing technologies being used.

11.2.2.2 Real-time data processing and analytics

The importance of real-time data processing and analytics is clearly visible in Industry 4.0 environments due to the nature of timely decision requirements. Today, the ongoing research challenges are to find the algorithms and architectures capable of operating with large numbers of data streams at high speed with low latencies. Edge computing, where data are processed locally (near the edge), is a hot research area that helps alleviate the load on central data centers and execute data processing faster. Further, the integration of AI and machine learning models for real-time analytics is another area of concentration for improved predictive maintenance, quality control, and operational capability.

11.2.3 Cybersecurity

11.2.3.1 Protecting data and systems from cyber threats

Since Industry 4.0 focuses on the interconnectedness of systems, data and networks also need security enforcement. Effective security is essential to safeguard sensitive data and industrial control systems from cyber threats. Research in this area is directed towards the evolution of more sophisticated encryption techniques, intrusion detection systems, and cybersecurity frameworks specific to industrial settings. At the same time, work is progressing hard and fast on the development of resilient infrastructures able to resist and recover from cyberattacks that may compromise the continuity and the safety of operations.

11.2.3.2 Ensuring secure communication and transactions

Keeping secure communication and transactions are important aspects of Industry 4.0, where you have to make sure a data breach does not happen and data transfer is always in an intact way. Among the research challenges are secure communication protocols designed to function effectively in real-time, resource-limited settings. The issue of the secure and transparent transfers is expected to be resolved by exploring blockchain technology for the flow of commercial products – a blockchain offers a tamper-proof record that can improve trust and certainty in industrial operations.

11.2.4 Human-machine interaction

11.2.4.1 Enhancing user interfaces and experiences

The importance of Human-Machine Interaction (HMI) Touchstone Labs Research is also focused on developing easy-to-use and highly intuitive interfaces for any type of system, designed for optimized operator experience and productivity. Augmented and virtual reality (AR and VR) bring immersive and interactive experiences that have the potential to open up whole new worlds. This results in the use of adaptive interfaces, which are capable of modifying and personalizing the workflows based on the user's preferences to make the system simpler and more efficient.

11.2.4.2 Collaborative robots (cobots)

Collaborative robots, or cobots, are designed to work alongside humans, enhancing productivity and safety in industrial environments. Research challenges include developing advanced sensing and control algorithms that enable cobots to safely interact with human workers. Efforts are also focused on improving the flexibility and adaptability of cobots to handle a variety of tasks. Human-robot collaboration models are being studied to optimize workflows and ensure seamless integration of cobots into existing processes.

11.2.5 Sustainability and environmental impact

11.2.5.1 Reducing energy consumption

One of the main goals of Industry 4.0 focuses on energy savings to create a more sustainable – and cost-effective – environment. The research is focused on novel low-energy technological development and optimization of energy consumption during manufacturing. This can range from smart grids, energy management systems, and predictive maintenance to track and minimize their energy expenditure. Industrialization adds more to this list, like searching for renewable options for energy and incorporating them in the functioning of industries, which is a topic that undergoes studies to bring sustainable manufacturing practices in place.

11.2.5.2 Implementing sustainable manufacturing practices

Adopting sustainable manufacturing practices is necessary to reduce the environmental fallout of industrial undertakings. For the research challenges, that means creating processes and technologies that will result in less waste, less resource consumption, and less emissions. Sustainable value chains are emerging, driven by a variety of circular economy models that stress material recycling and reuse. In addition to these, life cycle assessment (LCA) tools are being established to control industries by estimating the environmental burden caused by their products and processes, configuring toward sustainable industrial development.

11.3 SECTOR-SPECIFIC RESEARCH QUESTIONS AND PROPOSITION

11.3.1 Manufacturing

11.3.1.1 Smart factories and automated production lines

- Research Questions:
    1. How can the factories be intelligently engineered to obtain the maximum efficiency and flexibility in production??
    2. How best to incorporate AI and IoT into automated production lines?
- Propositions:

1. Developing a modular architecture for smart factories to enable scalability and adaptability.
2. Performing real-time data analytics and implementing the machine learning algorithms can improve the process optimization and defect detection, finally resulting in higher product quality and less downtime.

11.3.1.2 Predictive maintenance

- Research Questions:
  1. Which machine learning models work best for predicting equipment breakdowns?
  2. Where should predictive maintenance systems fit in the continuum of your manufacturing process?
- Propositions:
  1. Leveraging sensor data and advanced analytics can predict equipment failures accurately, reducing unexpected downtime and maintenance costs.
  2. Integrating predictive maintenance with IoT platforms can provide continuous monitoring and automated alerts, ensuring timely maintenance actions.

11.3.2 Healthcare

11.3.2.1 Personalized medicine and smart healthcare systems

- Research Questions:
  1. How can data/class predictive models and AI be used with intake assessment data and case plans to create personalized treatment plans?
  2. How hard can it be to piece all the data from different parts of the healthcare environment together to create the whole of a smart healthcare system?
- Propositions:
  1. Using sensor data and advanced analytics, equipment breakdowns can be predicted with great accuracy, serving to minimize both unanticipated downtime and maintenance costs.
  2. The combination of predictive maintenance and IoT platforms offers real-time monitoring and warning accompanied by automatic alerts and has potential for real-time monitoring and warning, helping ensure that maintenance is done now it is required.

11.3.3 Logistics and supply chain

11.3.3.1 Real-time tracking and inventory management

- Research Questions:
  1. How to improve IoT and RFID technologies for real-time tracking and inventory management?
  2. What are the advantages and disadvantages of using blockchain for supply chain transparency?
- Propositions:
  1. By augmenting IoT sensors along with RFID technology, it enables the up-to-the-minute tracking of inventory status location. Thus, advancing the productivity of the supply chain, as well as lowering the losses.
  2. Blockchain technology can provide a secure, transparent, and tamper-evident history of transactions, resulting in traceability and accountability in supply chains.

11.3.3.2 Autonomous vehicles and drones

- Research Questions:
  1. How are logistics compromised by safety and regulations when it comes to deploying driverless vehicles or drones?
  2. In what ways can AI help optimize self-driving delivery systems to make them more efficient and reliable?
- Propositions:
  1. For widespread use in logistics, autonomous vehicles and drones have to develop strong safety protocols and obtain regulatory approvals.
  2. AI-driven route optimization and dynamic scheduling can make autonomous delivery systems more efficient and reliable, thereby reducing costs and delivery times.

11.3.4 Energy

11.3.4.1 Smart grids and energy management systems

- Research Questions:
  1. What can be done to create smart grids to integrate renewable energy sources more effectively?
  2. What are the best practices for executing energy management systems in industrial settings?
- Propositions:
  1. Advanced control systems combined with real-time data analytics provide the key to balancing supply and demand and

enabling optimal integration of renewable energy sources in the smart grid.

2. Using artificial intelligence and IoT-based energy management systems can further enable efficiency in energy consumption, savings in cost, and renewed sustainability in industrial operations.

11.3.4.2 Renewable energy integration

- Research Questions:
    1. What is holding back clean energy integration on a massive scale because of technology and in underlying economic terms?
    2. How can the energy storage technologies be maximized and support renewable energy augmentations?
- Propositions:
    - Overcoming technology-specific barriers such as grid stability and storage capacity, which are essential for renewable energy integration at a large scale.
    - Fostering advanced energy storage solutions, such as high-capacity batteries and smart energy management systems, can improve the efficiency and reliability of renewable energy, assisting its assimilation into the grid.

11.4 FUTURE DIRECTIONS AND RESEARCH OPPORTUNITIES

11.4.1 Innovative technologies on the horizon

The pathways via leveraging quantum computing, AR and VR technologies, and 5G/6G technologies have the potential to revolutionize the fourth industrial revolution and thus create comprehensive research opportunities. What is a game-changer about quantum computing in manufacturing and logistics? There are a lot of complex optimization problems in manufacturing and logistics which can enhance efficiency and productivity on a large scale for optimization. Quantum algorithms to accelerate machine learning models target predictive maintenance and quality control while researchers can develop prototypes to compare performance with classical solutions. The focus of research might be on evaluating the performance of AR/VR on increasing the efficiency and safety of workers and developing immersive simulations for factory planning, layout, and process optimization. With the advent of 5G technology, the researchers will investigate the impact of 5G ultra-reliable low-latency communication (URLLC) on IoT device connectivity and real-time data processing in industrial environments. Smart factories can test integrated 5G technologies for real-world gains in productivity and data transmission reliability.

11.4.2 Collaborative research initiatives

Interdisciplinary and intersectoral cooperation are essential to tackle the multifaceted challenges of Industry 4. Experts from AI, robotics, materials science, and cybersecurity work together to create applications: this represents the spirit of interdisciplinary collaboration. Research should aim at collaborative and social science-incorporated projects to realize the human aspect of these technologies. The research consortia that include multiple universities, industry partners, and government agencies breed innovation because many different backgrounds are coming to the same table. In contrast, cross-sector collaboration is the study of proven ways that different industries team up to learn best practices and approaches. A key way of doing this is to forge inclusive, bias-free collaborative ecosystems for data sharing and collaboration around common problem-solving. Practical illustrations – such as smart city or healthcare logistics use cases – can act as pilot projects to demonstrate the possibilities and benefits of cross-sector collaboration, and thus increase the focus on broader adoption.

11.4.3 Long-term impact and vision

The potential for Industry 4.0 over the horizon and what its lasting impact could mean for the workforce, sustainability, and the larger economy. This knowledge can be used to inform how education curricula and training programs need to adapt to changes driven by Industry 4.0 in employment patterns, job roles, and skills. Tracking job market changes over time in longitudinal studies can generate new information about the evolution of the labor market and the success of educational programs. Sustainability-wise, researchers can look at whether circular economy principles can help to reduce waste and support more efficient use of resources within the context of Industry 4.0. This paper assesses how advanced manufacturing technologies may contribute to sustainability and investigates the scope and focus of present sustainability solutions by considering the life cycle environmental impact of ten relevant technologies. A global economic perspective on Industry 4.0 would then involve changes in trade patterns, supply chain dynamics, and economic inequality. This includes conducting economic impact studies, which can help inform public and private stakeholders about the potential opportunities and challenges of these technologies on a global basis, and an analysis of the role of policy and regulation in determining the adoption and impact of these technologies. These research efforts contribute to a better understanding of the key areas where concentrated efforts will lead to the successful adoption and deployment of Industry 4.0 technologies for sustainable growth and prosperity.

11.5 CONCLUSION

The fourth industrial revolution, under the name of the Industry 4.0 era, introduces multiple emerging technologies that include quantum computing, AR, VR, 5G, and a bigger, faster, and more interconnected network for what we already know in the industry. The promise of these disruptive technologies is enormous. Quantum computing may turn industry problem-solving in manufacturing and logistics on its head and lead to exponential improvements in efficiency. Workplace Safety and Productivity AR and VR can revolutionize training and remote assistance, making the workplace safer and more productive. At the same time, 5G technology should provide a more responsive, capable link for IoT devices to deliver real-time data processing and automation, which will be key for the smart factory of the future. However, a range of research challenges need to be tackled if we are to fully exploit the promise of Industry 4.0 through collaborative initiatives. Cross-collaboration among the disciplines will help to combine the expertise in AI, robotics, materials science, and cybersecurity, making it more coherent and complete in solving challenging industrial issues. Including social sciences in these initiatives may help to make sure that Industry 4.0 technologies are conceived within a broader understanding of the human consequences, ultimately helping to ensure more equitable access. By forging research consortia that link universities, industry partners, and government agencies, innovation can be catalyzed through the dispensation of shared information and combined efforts. Collaboration Matters Across Sectors, Too They act as case studies showing how best practices should be and the kind of frameworks that work and a culture of open working and transparency that is encouraged. Industries can crack this silo envied by building ecosystems around data sharing and combining problem-solving. This cooperation can lead to more integrated and specialized services, as demonstrated in the case of pilot projects running in sectors such as smart cities and healthcare logistics, which can then spread adoption and cross-sector cross-pollination.

In the future, the enduring effects of Industry 4.0 on the workforce, sustainability, and the worldwide economy are possibilities and risks alike. This requires research to understand the changing employment and job roles and skills due to the Industry 4.0 technologies even before they begin entering the workforce. It is imperative to create educational curricula and training programs to have a well-prepared workforce. Policymakers and educators alike might gain insights from longitudinal studies tracking changes in the job market over time and the effectiveness of education programs, ensuring the workforce reflects the demands of the future. Another essential focus is sustainability. It can support green industrialization by examining how to lower waste and improve resource productivity, which in research to circular economy principles reduce waste from industry and improve resource productivity. Understanding the long-term environmental implications of advanced manufacturing technologies enables a more complete approach to sustainability frameworks and metrics for application in different industrial systems. Smart grids and advanced energy management systems will also open new doors to fully integrate renewables, only strengthening Industry 4.0's pivotal role in environmental care. And finally, why it is so vital to grasp the global economic impact of Industry 4.0? Policies and regulations that affect the diffusion and impact of these technologies can be improved based on analysis of shifts in trade patterns, supply chain dynamics, and economic inequality. Economic impact studies act as a main source of information for policymakers and stakeholders to design strategies for capturing the opportunities Industry 4.0 offers with the least social costs.

REFERENCES

Adeoye, Ibrahim. 2024. "Fortifying Retail: Blockchain's Sentinel Role in Supply Chain Integrity, Transaction Security, and Data Safeguarding." http://dx.doi.org/10.2139/ssrn.4731687

Anthony Jnr, Bokolo. 2024. "Enhancing Blockchain Interoperability and Intraoperability Capabilities in Collaborative Enterprise-a Standardized Architecture Perspective." *Enterprise Information Systems* 18 (3): 2296647. https://doi.org/10.1080/17517575.2023.2296647.

He, Hongmei, John Gray, Angelo Cangelosi, Qinggang Meng, T. Martin McGinnity, and Jorn Mehnen. 2022. "The Challenges and Opportunities of Human-Centered AI for Trustworthy Robots and Autonomous Systems." *IEEE Transactions on Cognitive and Developmental Systems* 14 (4): 1398–1412. https://doi.org/10.1109/TCDS.2021.3132282.

Javaid, Mohd, Abid Haleem, Ravi Pratap Singh, Rajiv Suman, and Ernesto Santibañez Gonzalez. 2022. "Understanding the Adoption of Industry 4.0 Technologies in Improving Environmental Sustainability." *Sustainable Operations and Computers* 3: 203–17. https://doi.org/10.1016/j.susoc.2022.01.008.

Koh, Lenny, Guido Orzes, and Fu (Jeff) Jia. 2019. "The Fourth Industrial Revolution (Industry 4.0): Technologies Disruption on Operations and Supply Chain Management." *International Journal of Operations & Production Management* 39 (6/7/8): 817–28. https://doi.org/10.1108/IJOPM-08-2019-788.

Majid, Mamoona, Shaista Habib, Abdul Rehman Javed, Muhammad Rizwan, Gautam Srivastava, Thippa Reddy Gadekallu, and Jerry Chun-Wei Lin. 2022. "Applications of Wireless Sensor Networks and Internet of Things Frameworks in the Industry Revolution 4.0: A Systematic Literature Review." *Sensors* 22 (6): 2087. https://doi.org/10.3390/s22062087.

Mourtzis, Dimitris, John Angelopoulos, and Nikos Panopoulos. 2022. "Digital Manufacturing." In *The Digital Supply Chain*, 27–45. Elsevier. https://doi.org/10.1016/B978-0-323-91614-1.00002-2.

Tran, Park, Nguyen, and Hoang. 2019. "Development of a Smart Cyber-Physical Manufacturing System in the Industry 4.0 Context." *Applied Sciences* 9 (16): 3325. https://doi.org/10.3390/app9163325.

Vinodh, S, and Vishal Ashok Wankhede. 2021. "Application of Fuzzy DEMATEL and Fuzzy CODAS for Analysis of Workforce Attributes Pertaining to Industry 4.0: A Case Study." *International Journal of Quality & Reliability Management* 38 (8): 1695–1721. https://doi.org/10.1108/IJQRM-09-2020-0322.

Wankhede, Vishal Ashok, and S Vinodh. 2022. "State of the Art Review on Industry 4.0 in Manufacturing with the Focus on Automotive Sector." *International Journal of Lean Six Sigma* 13 (3): 692–732. https://doi.org/10.1108/IJLSS-05-2021-0101.

Xiong, Yi, Yunlong Tang, Qi Zhou, Yongsheng Ma, and David W. Rosen. 2022. "Intelligent Additive Manufacturing and Design: State of the Art and Future Perspectives." *Additive Manufacturing* 59 (November): 103139. https://doi.org/10.1016/j.addma.2022.103139.

Xu, Li Da, Eric L. Xu, and Ling Li. 2018. "Industry 4.0: State of the Art and Future Trends." *International Journal of Production Research* 56 (8): 2941–62. https://doi.org/10.1080/00207543.2018.1444806.

Index

Accuracy, 32, 48, 71, 124, 128, 130–131, 139–140, 152, 158, 163, 167
Adaptive control systems, 58–59, 62
Adaptive interfaces, 165
Additive manufacturing (AM), 13, 24, 30, 52–53, 89, 97, 122, 125, 131, 163, 173
Advanced analytics, 15, 32, 56, 122, 130, 134, 167
Advanced robotics, 2, 12, 35
Advanced sensing, 166
Aerospace, 13, 61, 68, 88
Agility, 9, 17, 29, 32, 46, 89, 124–126
Agriculture, 66–67
AI-Driven Insights, 68, 163
Airbus, 2
Artificial intelligence (AI), 2, 28, 45, 48, 51, 54, 90, 106, 111, 122–123, 131, 136, 159–160, 169
Asset tracking, 123
Augmented reality (AR), 6, 21, 23, 27, 30, 45, 70, 88–89, 125, 131
Automation, 2, 4–6, 8, 10, 15–16, 20, 23–24, 28, 31, 33–34, 36–37, 42–52, 55, 58, 61, 64, 84–85, 87–88, 94, 111, 119, 122, 124, 129, 131, 160–163, 171
Autonomous decision-making, 12, 54, 59, 62, 124
Autonomous systems, 47, 111, 119, 172
Autonomous vehicles, 55, 168

Backorder prediction, 137–140, 158–159
Backorder scenario, 139, 158
Backorder status, 141
Big data analysis, 89
Big data analytics, 1, 6–7, 9, 12, 14, 19, 36, 48, 54, 56, 62, 98, 105, 109, 112–113, 124, 129, 134–136, 160, 162
Big data processing, 24
Blockchain, 13–14, 20–21, 29, 46, 56, 60, 62–63, 126, 129, 132, 160, 165, 168, 172
Blockchain technologies, 13–14
Blockchain technology, 62, 132, 165, 168
Bosch, 2, 5
Bottleneck, 15, 95
Boxplot, 143–145
Business models, 4–5, 8–9, 21, 126, 160, 163

Circular economy, 13–14, 16, 20–22, 29, 38, 64, 111, 166, 170
Clean technologies, 113
Cloud-based platform, 1, 111
Cloud computing, 6, 8, 23–24, 27–28, 32, 37, 42, 56, 62, 86, 89, 93, 111, 118, 125, 131
Cloud platforms, 23, 71, 75–76, 164
Cloud services, 3, 70
Cognitive load, 161
Cognitive technologies, 93
Collaborative robots (Cobots), 15, 161, 165
Competitiveness, 20, 32, 48–49, 54, 122, 162–163
Competitive priorities, 24
Condition monitoring, 86
Connected robots, 1
Connectivity, 2–3, 18, 24, 29, 31, 44, 48, 54, 60, 62, 73, 77–78, 161
Context-aware computing, 57
Control algorithms, 36, 42, 55, 166
Co-robots, 59
Corporate social responsibility, 25, 112
Cost, 1, 4–5, 13, 16, 20, 23–25, 28, 31, 35–36, 47, 49, 51, 56, 58, 60, 68, 100, 106–107, 109, 123–128, 135, 137–138, 161, 163, 166–169, 172
Customer-centric, 24
Customer satisfaction, 126–128, 137
Customization, 4–5, 13, 16, 23, 49, 125, 160, 163
Cyber-physical systems (CPS), 111, 160
Cybersecurity, 21, 31–32, 43, 45, 60–61, 89, 114, 161, 163, 165, 170–171

Data acquisition, 32, 42–43, 50, 58, 69–71, 73, 75, 82, 88
Data analytics, 1–4, 6–10, 12–15, 19, 23, 35–36, 48–49, 54, 56, 60, 62, 93, 98, 101, 105–106, 109, 112–113, 124, 129, 134–136, 160–162, 167–168
Data analytics tools, 161
Data compression, 164
Data-driven decision-making, 13, 31, 94, 97, 107, 123, 126
Data-driven production, 2, 90
Data exchange, 3, 8, 12, 18, 24, 58, 78, 122, 128, 132, 160
Data management, 3, 7, 18, 28, 75–76, 100, 113, 161, 164
Data privacy, 109–110, 112, 115–118, 120, 162–163
Data protection, 17, 110, 112, 114, 116, 163
Data storage systems, 76
Data streams, 131, 164
Data synergy, 56
Data transfer, 75, 165
Data visualization, 73, 78, 83, 132
Decision elements, 25
Decision-making framework, 92, 120
Decision-making processes, 7, 89, 92–94
Decision support systems, 93, 98
Defect rate, 1
Demand forecasting, 124, 130, 137
Detection, 68, 101, 106, 108, 125, 145, 165, 167
Deviation, 14–15, 66, 95, 138, 146
Digital representation, 128
Digital shadow, 65
Digital skillset, 163
Digital supply chain, 135–136, 172
Digital transformation, 8, 18–19, 21, 40, 46, 63, 103, 121–123, 134–135, 163
Digital twin, 2, 4, 13, 22, 35, 37, 40, 43, 45, 49, 65–69, 71–73, 75, 77, 79, 81, 83–88, 114, 116
Digital twin for service, 69
Distributed databases, 164
Dynamic resource allocation, 58

Eco-friendly products, 162
Economic inequality, 163, 170, 172
Edge computing, 35, 76, 164
Efficiency, 1–2, 6–8, 12–16, 20–21, 23–25, 27–29, 31–32, 35–36, 43, 46, 48–50, 55–56, 58–59, 61–62, 65, 68–69, 71, 73, 76, 84, 95, 98–101, 108–109, 111, 115, 122–124, 126–128, 130, 132–134, 158, 160–162, 164, 166, 169, 171
Emission reduction, 111
Emissions trading system (ETS), 114
Encryption techniques, 165
Energy efficiency, 29, 43, 59, 68–69, 73, 162
Energy-efficient technologies, 117, 163
Energy management systems, 166, 168–169, 172
Ensemble learning, 139, 151–152
Environmental impact assessment (EIA), 114
Environmental management systems, 117–118
Environmental regulations, 110, 114, 116–117
Environmental sustainability, 8, 172
Equipment health, 14
Ergonomics, 161

Financial sustainability, 112
Finished goods, 137
Fleet history, 65
Flexibility, 4–5, 9, 13, 16, 21, 24, 31–32, 35–36, 46, 49, 56, 58, 61–62, 95, 106, 124–126, 128, 163, 166
Flexible manufacturing systems, 24
Forecasting accuracy, 124, 128, 131
Fourth revolution, 1

GDPR compliance, 116
General electric, 1, 4
Global industries, 162–163
Gradient boosting, 139–140, 151, 158–159
Green CPS, 59–61, 64

Healthcare systems, 167
Human-centered design, 161
Human-machine collaboration, 3
Human-machine interaction, 5, 10, 58, 68, 84, 161, 165
Human-robot collaboration, 22, 30, 59, 61, 64, 166
Hybrid prototyping, 27

Industrial robotics, 58, 61
Industrial robots, 59
Industry 4.0, 1–24, 26, 28–30, 32, 34, 36–40, 42–56, 58, 60–64, 66, 68, 70, 72, 74, 76, 78, 80, 82, 84, 86–136, 138, 140, 142, 144, 146, 148, 150, 152, 154, 156, 158, 160–173
Industry 4.0 adoption, 19, 38
Industry 4.0 technologies, 7–9, 12–17, 19–21, 43, 50–51, 55, 62, 89–90, 92–93, 100, 108, 112, 119, 127–130, 132, 134, 136, 162, 170–172
Intellectual property (IP), 113
Intellectual property rights, 109, 113, 119, 121
Internet of Things (IoT), 1, 12, 23, 42, 54, 89, 111, 119, 122, 160

Interoperability, 4, 9, 16–17, 29, 32, 55, 60–61, 128–129, 131–132, 160–161, 163–164, 172
Intrusion detection systems, 165
Inventory levels, 128, 130, 138
Inventory management, 6, 123, 127–128, 137–141, 143, 145, 147, 149, 151, 153, 155, 157–159, 168
IoT sensors, 14, 31–32, 130, 168
IP management, 115

Just-in-time manufacturing, 163

Key performance indicators (KPIs), 94, 122
Kiva robots, 1

Lead time, 13, 47, 122, 125–127, 129–134, 138–139, 142
Lean manufacturing, 16, 22, 92, 107
Lean six sigma, 15–16, 135, 173
Legal compliance, 110, 118, 162
Legal framework, 118
Life cycle assessment (LCA), 111, 119, 166
Logistics, 1, 12, 29, 31–32, 38, 40, 93, 96, 107, 124–126, 130, 135, 158, 160, 168–169, 171

Machine learning, 1, 12, 27–28, 33–34, 36, 39, 47–48, 51, 56, 62, 85, 90, 94, 100, 102, 105, 111, 120, 124, 131, 136–137, 139–143, 145–147, 149–153, 155, 157–159, 161, 164, 167, 169
Machine learning applications, 137, 139, 141, 143, 145, 147, 149, 151, 153, 155, 157, 159
Maintenance optimization, 96, 105
Manufacturing, 1–90, 92–126, 128–132, 134–140, 142, 144, 146, 148, 150, 152, 154, 156, 158–160, 162–164, 166–173
Manufacturing sector, 9, 19, 23, 25, 46–47, 55, 92, 98, 119
Manufacturing technologies, 8, 18, 38–39, 106, 160, 172
Mass customization, 4, 13, 16, 163
Mass production, 2, 12, 109
Modular architecture, 167
Modularity, 4
Multiphysics, 65

National inventory, 145

On-demand production, 24, 125, 131
Online monitoring, 65, 85
Operational costs, 58, 123, 163
Operational efficiency, 6, 8, 25, 36, 56, 58, 84, 122, 127, 130, 161
Operational integration, 56
Outlier detection, 145

Policy research, 162
Predictive analytics, 56–57, 163
Predictive maintenance, 4, 8, 13, 16, 20, 25, 35–36, 49, 54, 56–57, 59, 61, 68, 71, 88, 104, 106–107, 111–112, 123–124, 161, 165–167, 169
Predix platform, 1
Preventive maintenance, 56, 65
Principal component analysis, 139
Process control, 16, 70, 90, 126
Process deviations, 14–15
Process efficiency, 13, 15, 56
Process improvement, 15–16, 120
Process optimization, 13, 15–16, 32, 35–36, 56, 62, 161, 169
Product avatar, 65, 85, 87
Production flexibility, 13, 61
Production process, 4, 13–16, 29, 31–32, 36, 42, 45, 48, 55–57, 66, 95, 123, 130–131, 137, 160
Productivity, 1–7, 12, 14–15, 20, 32, 36, 39, 42, 45–49, 54, 57, 62, 67, 72, 84, 99, 101, 109, 111, 117, 122, 124–127, 133, 160, 163, 165, 168–169, 171–172
Product lifecycle management, 14, 35, 54, 57, 85
Product personalization, 24
Protocols, 74–75, 77–78, 86, 161, 163–165, 168

Quality assurance, 13, 35
Quality control, 6, 35, 56, 62, 93, 95, 97–98, 102, 124, 165, 169
Quality control tools, 93, 97, 102

Raw materials, 31, 58, 71, 117, 137
Real-time analytics, 164
Real-time data, 1–2, 5, 8, 12, 16, 32, 35–36, 42, 54, 58–59, 73, 79, 83, 123, 125, 128, 130–132, 164, 167–169, 171
Real-time data processing, 164, 169
Real-time decisions, 31, 161
Real-time insights, 25, 30, 130
Real-time monitoring, 8, 13–14, 27, 56, 59–60, 62, 65–66, 71, 73, 76, 83, 111–112, 123, 126, 167
Real-time tracking, 57, 163, 168
Recurrent neural network, 158
Recyclability, 14
Regulatory compliance, 17, 47, 109, 114
Regulatory frameworks, 33, 118, 163
Remote monitoring, 28, 43, 67, 84

Replenishment, 138, 159
Research challenges, 160, 162–166, 171
Resilience, 46, 123–126, 128, 130, 132–136
Resilient, 36, 127, 130, 165
Resilient infrastructure, 165
Resource allocation, 4, 58, 128
Resource efficiency, 14, 20, 25, 28, 59, 111
Resource management, 5
Resource optimization, 124
Resource utilization, 122, 124, 126–128, 131, 133–134
RFID technologies, 168
Risk assessment, 96, 115, 122, 128–129, 131–133
Risk management, 26, 110, 115, 118, 121, 125–126
Robotics, 2, 5, 12, 14–15, 19, 24, 33, 35, 39, 46, 58, 61, 86, 89, 109, 111, 119, 122, 124, 129, 131, 133, 135, 161, 163, 170–171

Safety stock, 138
Scalability, 55–56, 62, 70, 125, 167
Secure communication, 165
Sensor data, 65, 76, 167
Siemens, 1, 6, 67, 115–117, 120
Simulation, 13, 26, 35, 45, 63, 65–66, 68–69, 72–73, 86–87, 93, 95, 98, 105, 128, 131, 135, 158, 169
Simulation and training, 68
Simulation technology, 68–69
Simulation tools, 35
Small and medium-sized enterprises (SMEs), 21, 38
Smart data analytics, 93
Smart factories, 5, 12, 25–26, 31, 36, 109, 111, 161, 163, 166–167, 169
Smart factory development, 25
Smart farming, 67
Smart grids, 43, 166, 168, 172
Smart manufacturing, 1, 5, 10, 14, 20–31, 33, 35–42, 52, 54, 57, 62–63, 86–88, 93, 99, 107, 116, 123, 135–136, 163
Smart performance, 24
Smart sensors, 48
Smart technologies, 9, 122
SMOTE, 139–140, 142, 148–149, 157–159
Social responsibility, 25, 112, 120
Storage costs, 138
Strategic direction, 25
Supply chain(s), 1–2, 6, 8, 10, 14, 16, 24, 26, 29, 31, 35, 36, 39, 48–51, 54, 56–57, 60–61, 66, 89, 93, 95, 99, 111–112, 117, 120, 122–128, 130, 132–133, 135–139, 158–159, 161, 163, 168, 170, 172
Supply chain management, 10, 29, 35, 51, 95, 130, 158–159, 172
Supply chain optimization, 8, 26, 57, 161
Sustainability, 1, 5, 8–11, 16, 20, 29–30, 32, 34–35, 37–39, 43, 50, 59, 61, 88, 98–99, 101, 106–107, 109–112, 114–119, 125, 132, 135, 162–163, 166, 169–172
Sustainable practices, 8–9, 109, 116, 124
Sustainable supply chains, 60–61
Synchronization, 23, 69, 71–72, 85, 93
System integration, 24, 45

Technology integration, 3, 31
Technology selection, 91, 104
3D printing, 46, 53, 131
Time to recovery, 127, 130
Traceability, 14, 42, 56, 60, 126, 132, 168
Transparent, 132, 165, 168

Univariate analysis, 142
Upskilling, 163

Virtual factory, 85
Virtual reality, 6, 23, 27, 69, 131, 165
Virtual replica, 65, 114
Virtualization, 4, 29, 128–129, 131
Visibility, 4, 14, 32, 57, 66, 115, 123–126, 132

Waste minimization, 162
Waste reduction, 15, 60

For Product Safety Concerns and Information please contact our EU representative GPSR@taylorandfrancis.com Taylor & Francis Verlag GmbH, Kaufingerstraße 24, 80331 München, Germany

Batch number: 10397790

Printed by Printforce, the Netherlands